Protein-Protein and Domain-Domain Interactions

Pandjassarame Kangueane
Christina Nilofer

Protein-Protein and Domain-Domain Interactions

Springer

Pandjassarame Kangueane
Biomedical Informatics
Pondicherry, India

Christina Nilofer
Biomedical Informatics
Pondicherry, India

School of Bio Sciences and Technology
VIT University
Vellore, Tamil Nadu, India

ISBN 978-981-10-7346-5 ISBN 978-981-10-7347-2 (eBook)
https://doi.org/10.1007/978-981-10-7347-2

Library of Congress Control Number: 2017964603

Printed on acid-free paper

This Springer imprint is published by Springer Nature
The registered company is Springer Nature Singapore Pte Ltd.
The registered company address is: 152 Beach Road, #21-01/04 Gateway East, Singapore 189721, Singapore

This book is dedicated to J. Janin, C. Chothia, and R. Nussinov for their scientific contribution in this field and to P. Balaram, C. Ramakrishnan, and P. Gautam for their personal support during this long journey through decades of excitement.

Preface

Protein complexes are often involved in catalysis, regulation, assembly, immunity, and inhibition. This involves the formation of stable interfaces between the interacting proteins. Therefore, it is important to understand the molecular principles of protein-protein binding. This book introduces the concepts in the understanding of protein-protein interaction (PPI) using several illustrations and examples. It illustrates the importance and significance of the molecular (physical and chemical) and evolutionary (gene fusion) principles of protein-protein and domain-domain interactions toward the understanding of cell division, disease mechanism, and target definition in drug discovery. It describes several issues associated with PPI using cutting-edge advancement in Bioinformatics through Bioinformation Discovery. The chapters provide information pertaining to the types of protein-protein complexes (homodimers, heterodimers, multimer complexes) in context with various specific biological functions. The significance of such complex formation in biology in the light of molecular evolution is also highlighted using examples. Hence, this book helps to introduce concepts in protein-protein binding in a step-by-step manner using examples, illustrations, and exercises for an easy understanding of the subject by undergraduate, graduate, and research students.

Pondicherry, India | Pandjassarame Kangueane

Acknowledgment

We wish to express our sincere appreciation to all members of Biomedical Informatics (India) and several research group members at MBU at IISc (India), CBT at Anna University (India), Bioinformatics Centre (BIC) at the National University of Singapore (Singapore), MAE at Nanyang Technological University (Singapore), VIT University (India), AIMST University (Malaysia), and RCSB (USA) for an extensive support on the subject of this book during the past decades.

We also thank Dr. Paul Shapshak (USA), Dr. Francesco Chiappelli (USA), Dr. Darren Flower (UK), Dr. Venkatrajan Mathura (USA), Dr. Subramanian Subbiah (USA), Dr. Tan Tin Wee (Singapore), Dr. Meena Sakharkar (Canada), Dr. Arumugam Mohanapriya (India), Dr. Sajitha Lulu (India), Dr. Prasanna R. Kolatkar (UAE), Dr. Gopichandran Sowmya (Australia), Dr. Velmurugan Karthikraja (USA), Dr. Abishek Suresh (Australia), Dr. Kannan Gunasekaran (USA), Dr. Badireddi Subathra Lakshmi (India), Dr. Ren Ee Chee (Singapore), Ms. Gunasegaran Shamini (Malaysia), and Ms. Sathyanarayanan Anitha (India) for fruitful discussions and knowledge sharing on several aspects of this subject.

Our thanks are also to reviewers, editors, readers with critical feedback, open-access movement (especially PubMed Central (PMC)), and many other publishers for their support on the subject of this book towards its specific advancement.

Contents

About the Authors

Pandjassarame Kangueane is a Director of Biomedical Informatics (P) Ltd., India, Chief Editor, Bioinformation, India, and Associate Editor, BMC Bioinformatics, UK. He has earlier served as Professor of Biotechnology at AIMST University, Malaysia (2009–2011), Visiting Professor of Bioinformatics at VIT University, India (2007–2009), Assistant Professor of Bioinformatics at Nanyang Technological University, Singapore (2002–2006), Bioinformatics Visiting Scientist at Chiron Corporation, USA (2001), and Bioinformatics Scientist at S*Bio Pte Ltd., Singapore (2001). His research interest is in Agriculture, Bioinformatics, and Biotechnology with specialization in peptide vaccine design, protein-protein interactions, enzyme engineering, genome analysis, and Agro-Informatics. He has been conferred with various prestigious awards, notably, Indian Leadership Award for Industrial Development by All India Achievers Foundation, Vishal Bharathi award with the title "Bharath Jothi" by GOPIO, France, and Ambassador for Peace by Universal Peace Federation for Contribution in Science and Technology. He has served as referee for a number of international journals, including *Bioinformatics* (UK), *Nucleic Acids Research* (UK), *BMC Genomics* (UK), and *BMC Bioinformatics* (UK). He has more than 15 years of teaching experience in Computational Biology, Bioprocess, Biotechnology, Computing, Molecular Modeling and Drug Design, Comparative Genomics, and Bioinformatics. He has also published more than 50 research articles in the peer-reviewed international journals and authored or co-authored and edited numerous books and book chapters. He is a member of many international scientific societies and organizations.

Christina Nilofer is a Scientist at Biomedical Informatics (P) Ltd. She received her Master of Science in Bioinformatics from Stella Maris College, Madras University, Chennai, India. She is a doctoral candidate at VIT University, Vellore, India. She has earlier served as a research member in Biotechnology (Saksin Lifesciences (P) Ltd.) and Bioinformatics (Genome Life Sciences (P) Ltd.) industries. She is an author of two book chapters (*Global Virology II*) and an article in a peer-reviewed journal.

Introduction

1

1.1 Proteins

Protein macromolecules are polypeptides made of amino acids (Fig. 1.1). A peptide bond covalently connects two amino acid residues (Fig. 1.2). There are 20 naturally occurring amino acids (Table 1.1). Amino acids are made of backbone and side chains (Fig. 1.1). They are broadly grouped into ten non-polar and ten polar amino acids (Fig. 1.3) based on their side chain features. They are further grouped into several subgroups based on chemical properties (Fig. 1.3).

1.2 Covalent Bond and Bond Length

Proteins are made of amino acid residues, and residues are made of atoms such as C, H, N, O, and S. These atoms are connected by covalent bonds (Fig. 1.4) in residues making proteins. Therefore, it is important to recollect such information in this context as shown in Fig. 1.5.

1.3 Disulfide Bond

A disulfide bond (S-S) is formed between two cysteine residues during the formation of the oxidized form of a cystine dimer (Fig. 1.6), and they play an important role in protein stability.

1.4 Protein Structure Representation

Proteins are represented using primary (Fig. 1.7), secondary (Fig. 1.8), tertiary (Fig. 1.9), and quaternary (Fig. 1.10) structures.

P. Kangueane, C. Nilofer, *Protein-Protein and Domain-Domain Interactions*,
https://doi.org/10.1007/978-981-10-7347-2_1

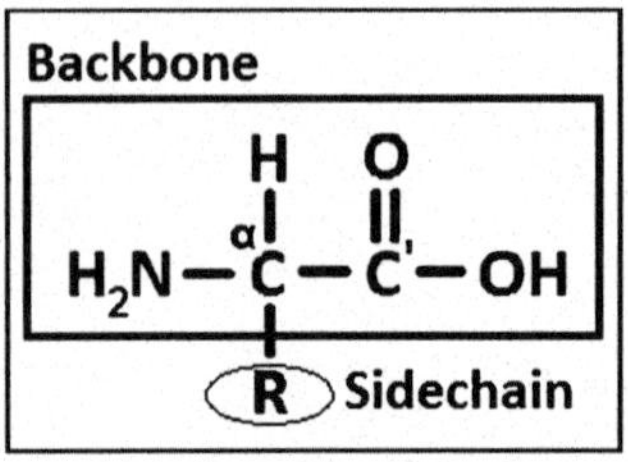

Fig. 1.1 Skeleton of an amino acid residue is depicted showing backbone and side chain. The side chain (R) varies for each of the 20 amino acids, while the backbone remains constant

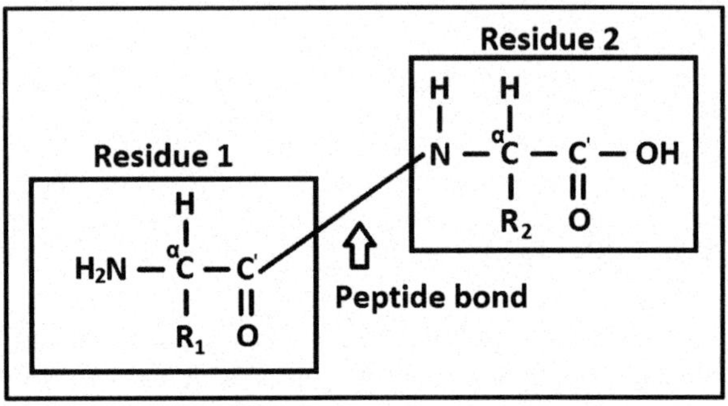

Fig. 1.2 A peptide bond is illustrated. Amino acid residues are linked by a peptide bond in proteins. A peptide bond is formed between the carboxyl group of one amino acid and the amino group of the other amino acid

Table 1.1 Three letter and single letter representations of amino acids with molecular weight

Amino acid name	Three letter code	Single letter code	Molecular weight
Alanine	Ala	A	89.1
Arginine	Arg	R	174.2
Asparagine	Asn	N	132.1
Aspartic acid	Asp	D	133.1
Cysteine	Cys	C	121.2
Glutamic acid	Glu	E	147.1
Glutamine	Gln	Q	146.1
Glycine	Gly	G	75.1
Histidine	His	H	155.2
Isoleucine	Ile	I	131.2
Leucine	Leu	L	131.2
Lysine	Lys	K	146.2
Methionine	Met	M	149.2
Phenylalanine	Phe	F	165.2
Proline	Pro	P	115.1
Serine	Ser	S	105.1
Threonine	Thr	T	119.1
Tryptophan	Trp	W	204.2
Tyrosine	Tyr	Y	181.2
Valine	Val	V	117.1

This data is collected from URL https://en.wikipedia.org under the open access creative commons attribution license

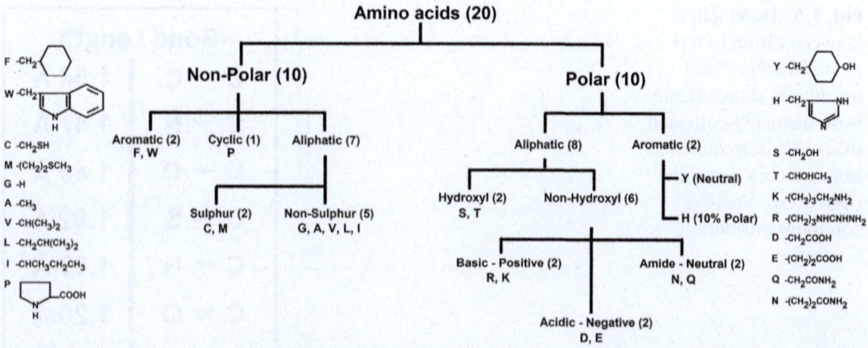

Fig. 1.3 Schematic classification of amino acids based on their physical and chemical properties. Refer to Table 1.1 for single letter abbreviation of amino acid residues. It is important to understand the physical and chemical features of amino acids to understand protein folding, structure, function, and interactions. The side chains for each amino acid are also illustrated using diagrams

Fig. 1.4 A covalent bond is illustrated using cysteine with corresponding bond length

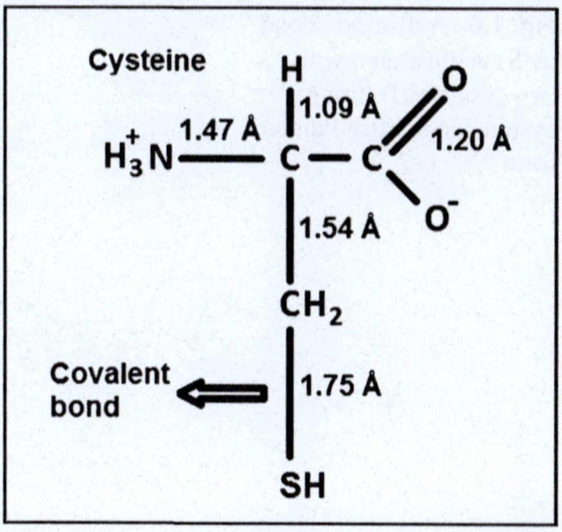

1.5 Structural Motifs

Proteins are made of domains and domains in turn are made of structural motifs. Different combinations of structural motifs are known. Examples of some structural motifs are illustrated in Fig. 1.11.

Fig. 1.5 Bond length between different types of atoms in amino acid residues is shown (Data were manually collected from URL http://chem.libretexts.org under the open access creative commons attribution license)

Bond Length	
C – C	1.54 Å
C – N	1.47 Å
C – O	1.43 Å
C – S	1.82 Å
C – H	1.09 Å
C = O	1.20 Å
C = C	1.34 Å
S – S	2.05 Å

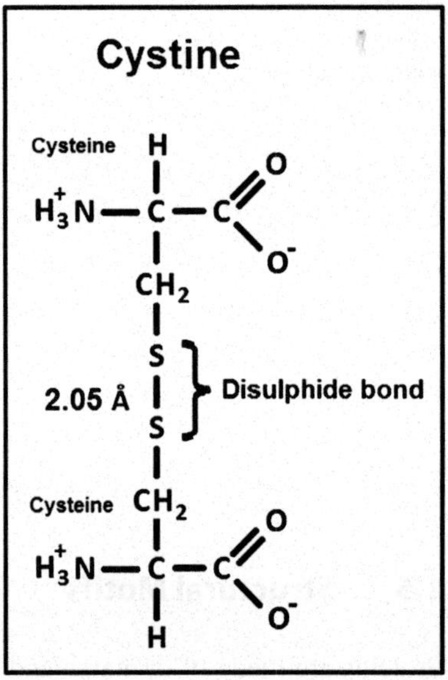

Fig. 1.6 A disulfide bond (S-S) is illustrated with two cysteines forming a cystine, an oxidized dimer form

1.6 Protein Surface and Core

The protein surface (accessible to water) and core (buried upon protein folding) is illustrated in Fig. 1.12.

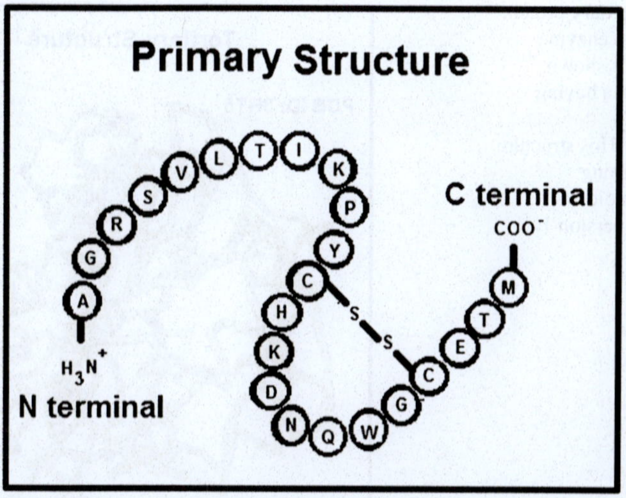

Fig. 1.7 Primary structure representation of a protein is shown. A series of amino acids linked together by a peptide bond forming the protein's primary structure is illustrated. N (amino) terminal and C (carboxyl) terminals are marked, and a disulfide bond between two cysteines is also shown

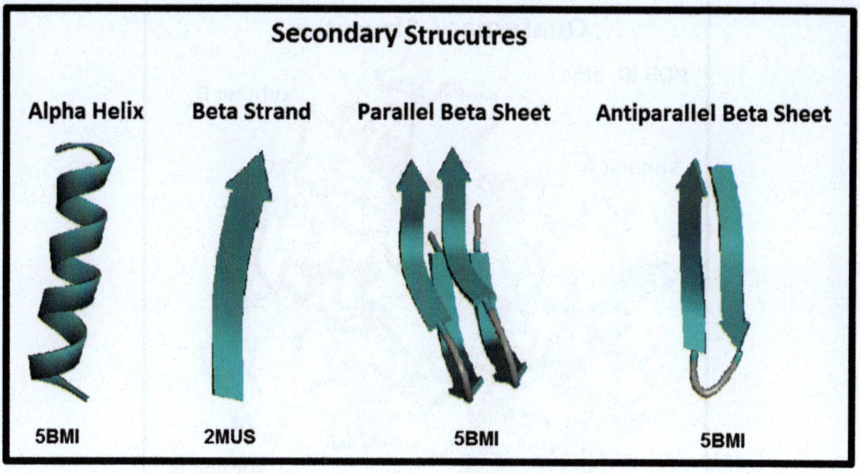

Fig. 1.8 Secondary structure representation of a protein is shown. Several forms of secondary structure elements (alpha helix, beta strand) are drawn using PDB [1] data having PDB ID (5BMI and 2MUS). Beta strands form beta sheets (parallel and antiparallel). The secondary elements were drawn using Discovery Studio Visualizer™ version 16.1 [3]. Please refer to other textbooks on protein structures for information on 3_{10} and π helix

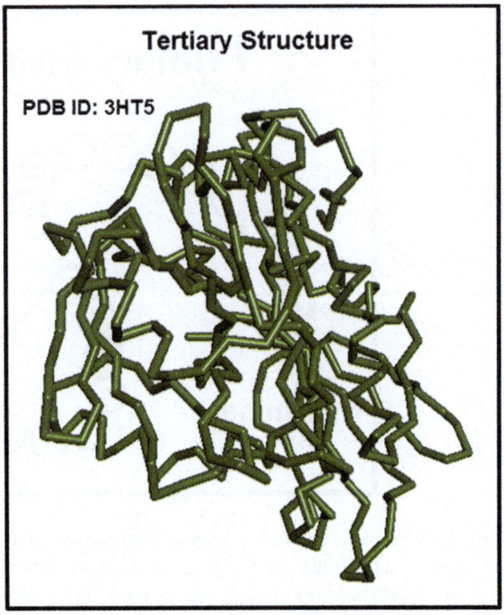

Fig. 1.9 A tertiary protein structure of an enzyme (transferase) is shown using PDB data having PDB ID 3HT5 (transferase). This structure is generated using Discovery Studio Visualizer™ version 16.1 [3]

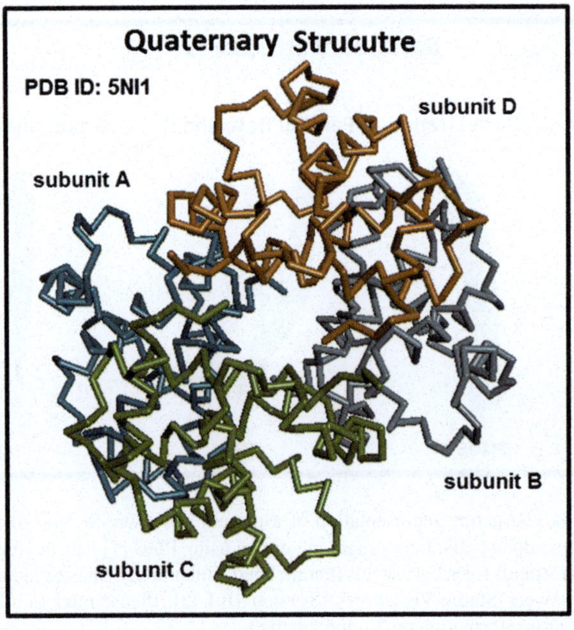

Fig. 1.10 A quaternary protein structure of hemoglobin is shown using PDB data having PDB ID 5NI1 (hemoglobin protein). This structure is generated using Discovery Studio Visualizer™ version 16.1 [3]

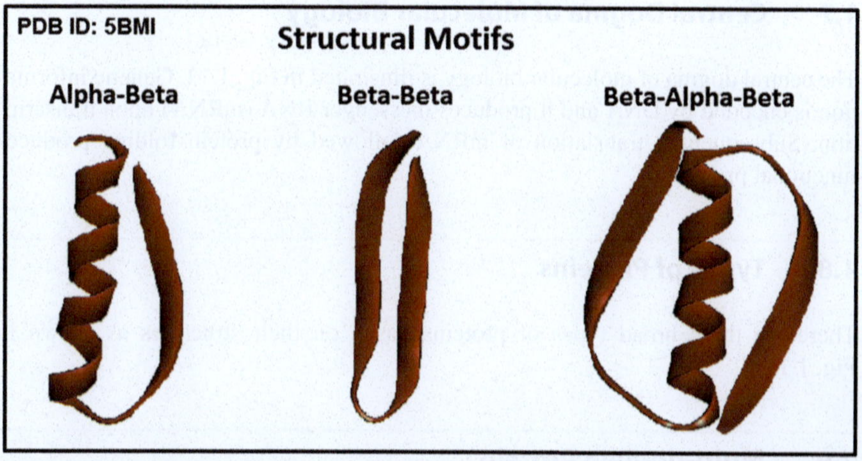

Fig. 1.11 Examples of some structural motifs (alpha-beta, beta-beta, and beta-alpha-beta) using PDB data having PDB ID: 5BMI are illustrated. This structure is generated using Discovery Studio Visualizer™ version 16.1 [3]

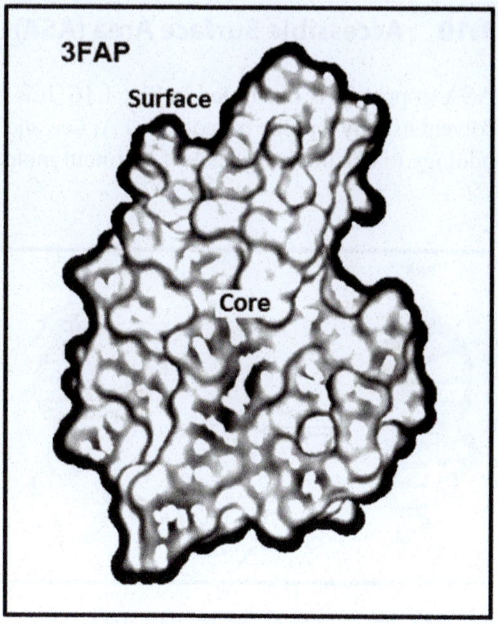

Fig. 1.12 Regions of protein (surface and core) are illustrated using PDB data having PDB ID: 3FAP (cell cycle protein complex) is shown. This structure is generated using Discovery Studio Visualizer™ version 16.1 [3]

1.7 Central Dogma of Molecular Biology

The central dogma of molecular biology is illustrated in Fig. 1.13. Genetic information is encoded by DNA and it produces messenger RNA (mRNA) upon transcription. Subsequently, translation of mRNA followed by protein folding produces functional proteins.

1.8 Types of Proteins

There are three broad types of proteins based on their structures as shown in Fig. 1.14.

1.9 Multi-domain Protein

A multi-domain protein is illustrated in Fig. 1.15. The alpha chain of human leukocyte antigen (HLA) is made of three domains, namely, alpha 1, alpha 2, and alpha 3.

1.10 Accessible Surface Area (ASA)

ASA of protein is illustrated in Fig. 1.16. It is the surface area made accessible by solvent usually a probe of radius 1.4 Å. Lee and Richard (1971) described the methodology to calculate the ASA of a protein molecule [5].

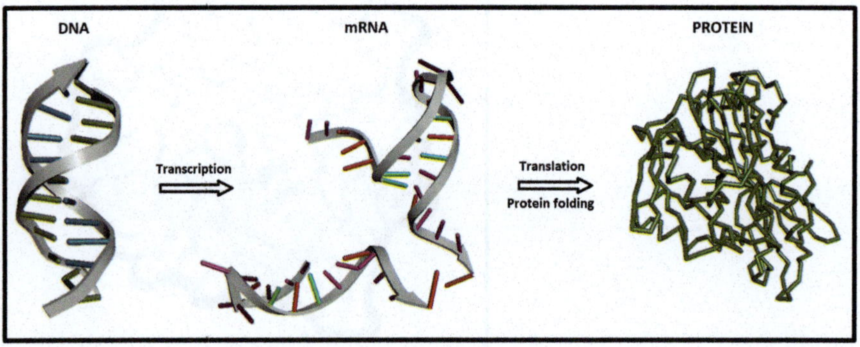

Fig. 1.13 The central dogma of molecular biology is shown schematically using PDB data for DNA (PDB ID – 1BNA), mRNA (PDB ID: 4I8R), and tertiary protein structure (PDB ID: 3HT5). This structure is generated using Discovery Studio Visualizer™ version 16.1 [3]

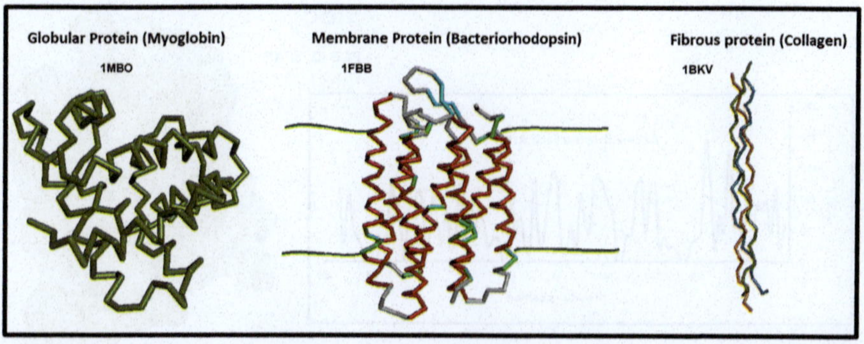

Fig. 1.14 Types of protein structures: globular protein (myoglobin) PDB ID: 1MBO, membrane protein (bacteriorhodopsin) PDB ID: 1FBB, and fibrous protein (collagen) PDB ID: 1BKV are illustrated. This structure is generated using Discovery Studio Visualizer™ version 16.1 [3]

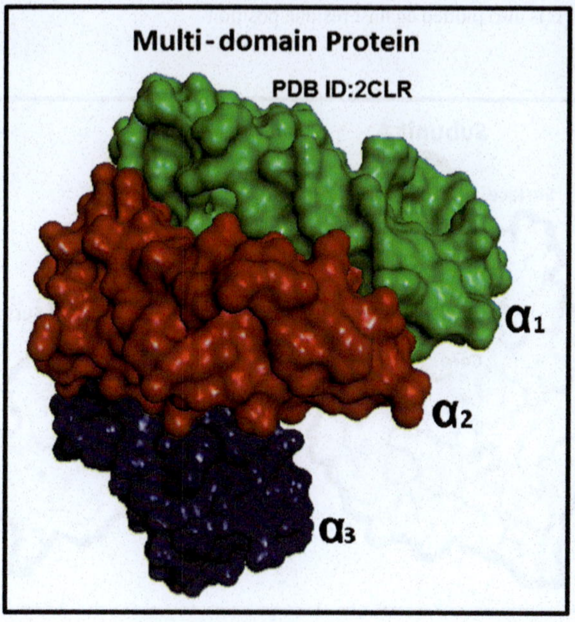

Fig. 1.15 A multi-domain protein is illustrated using PDB data having PDB ID: 2CLR (alpha chain of human leukocyte antigen). The alpha chain is made of alpha 1 (green), alpha 2 (red), and alpha 3 (blue) domains where the peptide-binding function groove is formed by alpha 1 and alpha 2. This structure is generated using Discovery Studio Visualizer™ version 16.1 [3]

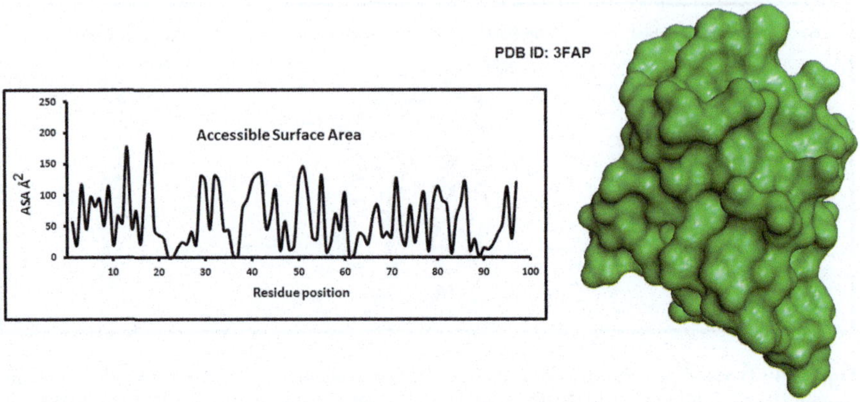

Fig. 1.16 Protein accessible surface area is shown for PDB ID: 3FAP (cell cycle protein subunit A) using SURFACE RACER™ [2]. This structure is generated using Discovery Studio Visualizer™ version 16.1 [3]. It is also plotted against residue position

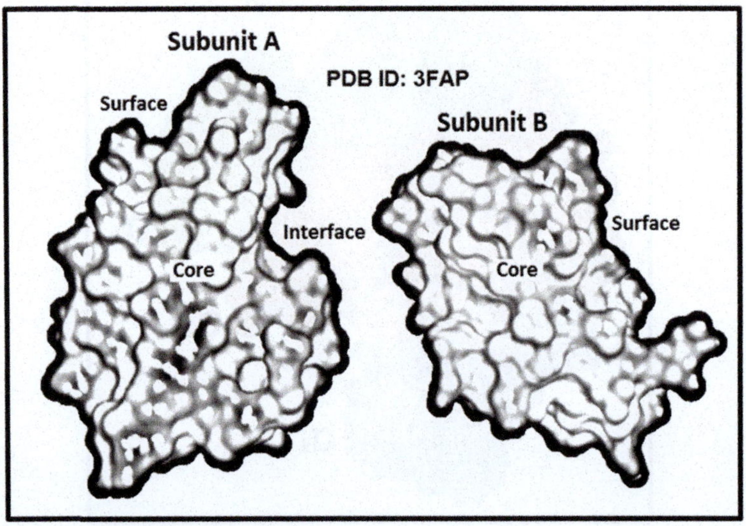

Fig. 1.17 Protein surface, core, and interface are shown in a complex using PDB ID: 3FAP using Discovery Studio Visualizer™ version 16.1 [3]

1.11 Protein-Protein Interface

A protein-protein interface is shown in Fig. 1.17. An interface is that part of the surface before complex formation that is involved in binding as shown in Fig. 1.18.

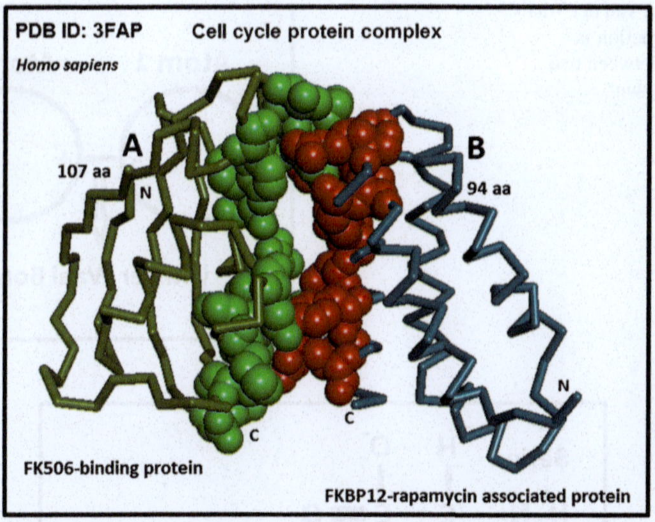

Fig. 1.18 An interface size (30 residues formed by 18 residues from subunit A and 12 residues from subunit B) is illustrated using PDB data having PDB ID: 3FAP (cell cycle protein complex). The interface residues are shown using CPK (Corey-Pauling-Koltun) model representation using Discovery Studio Visualizer™ version 16.1 [3]

$$\Delta \text{ASA (complex)} = \frac{\text{ASA (subunit A)} + \text{ASA (subunit B)} - \text{ASA (complex AB)}}{2}$$

Fig. 1.19 Interface area calculation is illustrated below using a formula

1.12 Interface Area

Interface area (ΔASA) is calculated using the formula described elsewhere [4] and as given below in Fig. 1.19.

1.13 Van Der Waals (vdW) Interactions

vdW (Fig. 1.20) is an important interaction at the protein-protein interface. Therefore, it is important to recollect information related this type of interaction.

1.14 Hydrogen Bonds

Hydrogen bonds (Fig. 1.21) are non-covalent interactions that are characteristics of protein-protein interfaces.

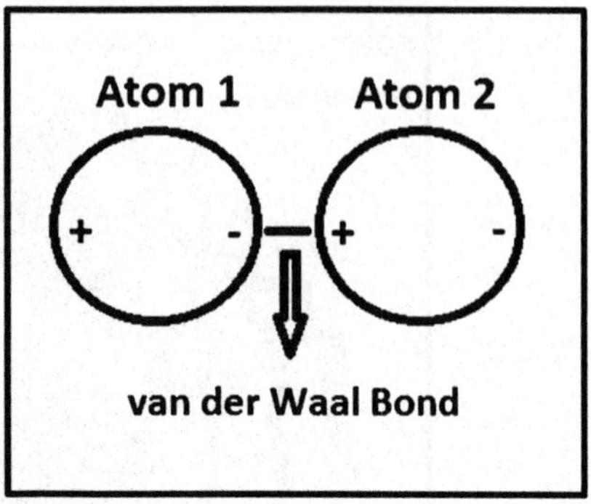

Fig. 1.20 A van der Waals (vdW) interaction is illustrated between two atoms as shown

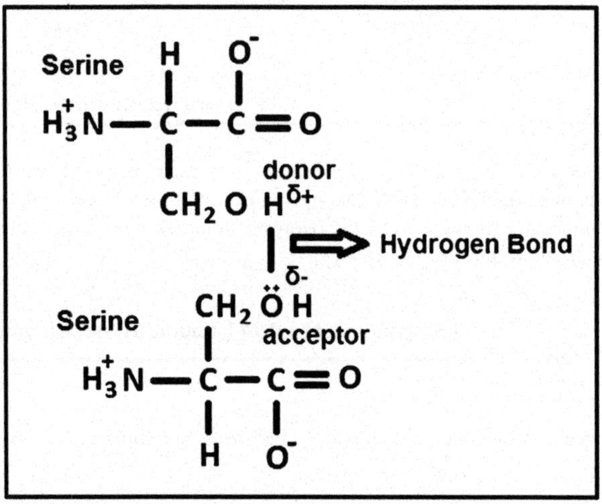

Fig. 1.21 A hydrogen bond between two serine residues is illustrated. An H bond is formed between a donor atom and an acceptor atom as shown

1.15 Salt Bridges

Non-covalent interactions at the protein-protein interface are characteristics of salt bridges in many complexes. Therefore, it is important to recollect information on salt bridge (Fig. 1.22).

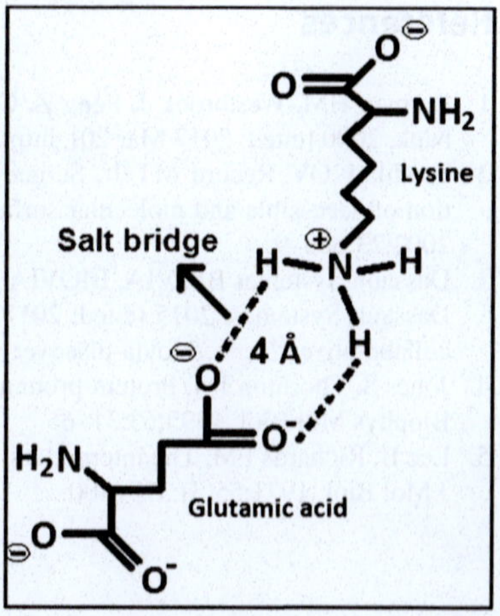

Fig. 1.22 A salt bridge is illustrated within a cutoff distance of 4 Å between glutamic acid and lysine residues

Exercises

1. Calculate the accessible surface area (ASA) for *Candida rugosa* lipase (PDB ID: 1CRL) using Surface Racer™, and plot its profile against residue position.
2. State the smallest amino acid residue.
3. Illustrate an alpha-beta-alpha motif using a neat diagram.
4. Illustrate a hydrogen bond with a neat diagram.
5. Illustrate a protein-protein interface with a labeled diagram.
6. Show core and surface regions in a protein structure.
7. State some non-covalent interactions.
8. Illustrate a salt bridge using a neat diagram.
9. Show secondary structure elements using diagrams.
10. Illustrate the calculation of interface area using an example.
11. State the heaviest amino acid.
12. State the longest amino acid.
13. Name any two programs to calculate accessible surface area (ASA) of proteins.
14. What are non-covalent interactions?
15. Illustrate bond length in a cysteine residue.
16. Show a disulfide bond using a neat diagram.
17. Illustrate a multi-domain protein using a diagram.
18. Show different regions of a protein-protein complex using a neat diagram.
19. Give a physical and chemical classification of amino acids using a chart.
20. Illustrate a globular, membranous, and fibrous protein using examples.

References

1. Berman HM, Westbrook J, Feng Z, Gilliland G, Bhat TN, Weissig H et al. The protein data bank. 2000 [cited: 2017 Mar 20]. http://www.rcsb.org/pdb/home/home.do
2. Tsodikov OV, Record MT Jr, Sergeev YV. Novel computer program for fast exact calculation of accessible and molecular surface areas and average surface curvature. J Comp Chem. 2002;23:600–9.
3. Dassault Systemes BIOVIA, BIOVIA Discovery Studio Visualizer, v16.1.0.15350, San Diego: Dassault Systemes; 2015 [cited: 2017 Mar 20]. Available from: http://accelrys.com/products/collaborative-science/biovia-discovery-studio/visualization-download.php
4. Jones S, Thornton JM. Protein-protein interactions: a review of protein dimer structures. Prog Biophys Mol Biol. 1995;63:31–65.
5. Lee B, Richards FM. The interpretation of protein structures: estimation of static accessibility. J Mol Biol. 1971;55(3):379–400.

Protein-Protein Binding

2

2.1 Molecular Functions of Protein Complexes

Protein complexes are associated with a number of molecular functions such as enzyme catalysis (Fig. 2.1), regulation (Fig. 2.2), biological assembly (Fig. 2.3), and immune response (Fig. 2.4). Protein-protein interaction is also seen among enzyme inhibitors (Fig. 2.5) and regulatory inhibitor (Fig. 2.6). This happens

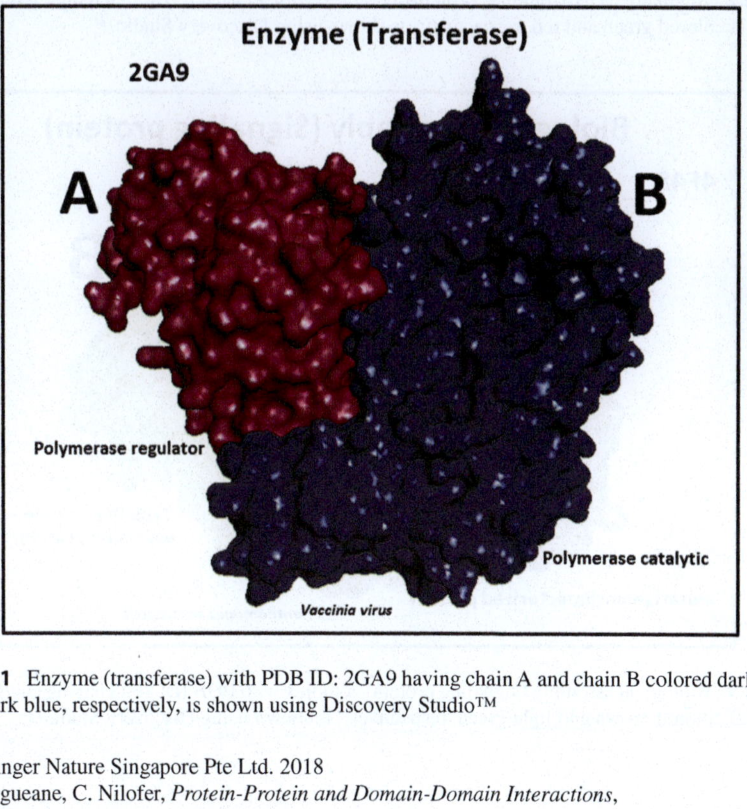

Fig. 2.1 Enzyme (transferase) with PDB ID: 2GA9 having chain A and chain B colored dark pink and dark blue, respectively, is shown using Discovery Studio™

P. Kangueane, C. Nilofer, *Protein-Protein and Domain-Domain Interactions*,
https://doi.org/10.1007/978-981-10-7347-2_2

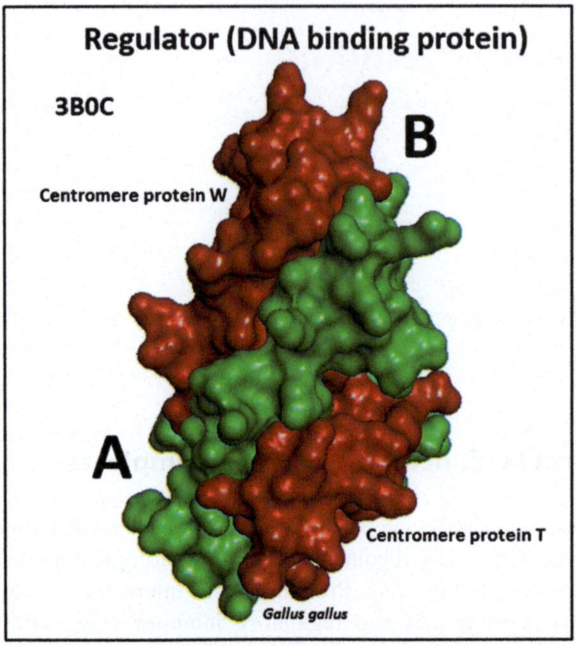

Fig. 2.2 Regulator (DNA-binding protein) complex with PDB ID: 3B0C having chain A and chain B colored green and red, respectively, is shown using Discovery Studio™

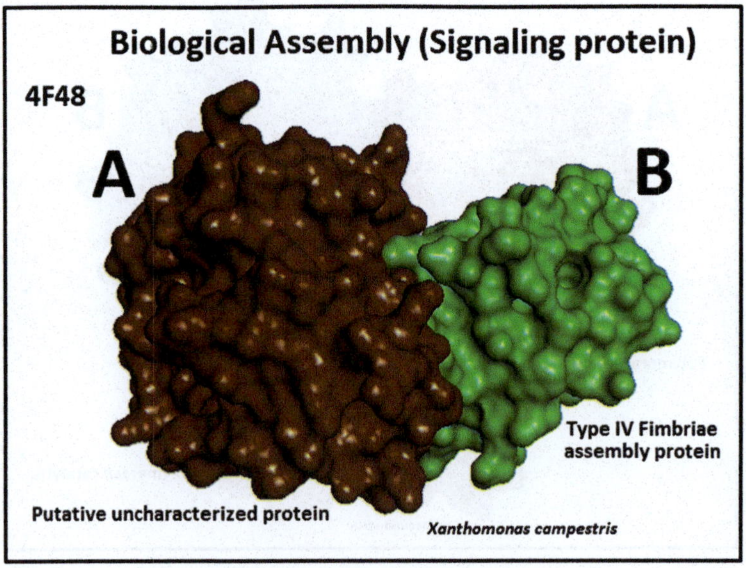

Fig. 2.3 Biological assembly (signaling protein) complex with PDB ID: 4F48 having chain A and chain B colored brown and light green, respectively, is shown using Discovery Studio™

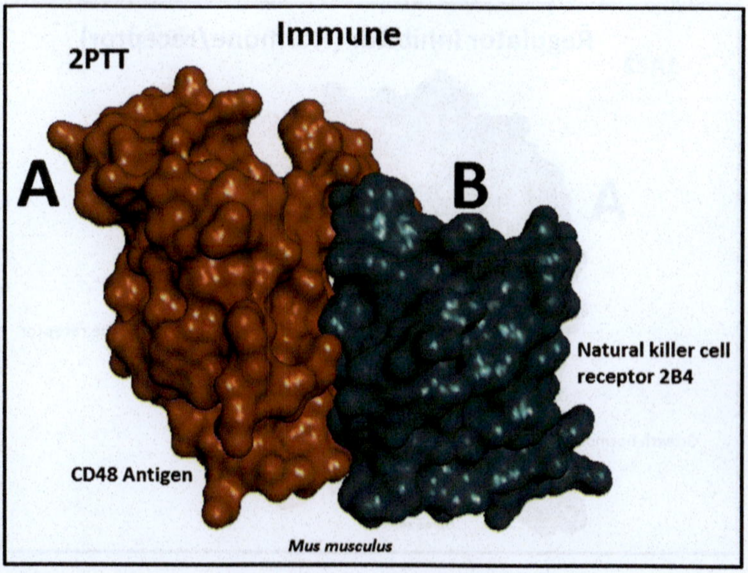

Fig. 2.4 Immune (PDB ID: 2PTT) complex having chain A and chain B colored with orange and cyan, respectively, is shown using Discovery Studio™

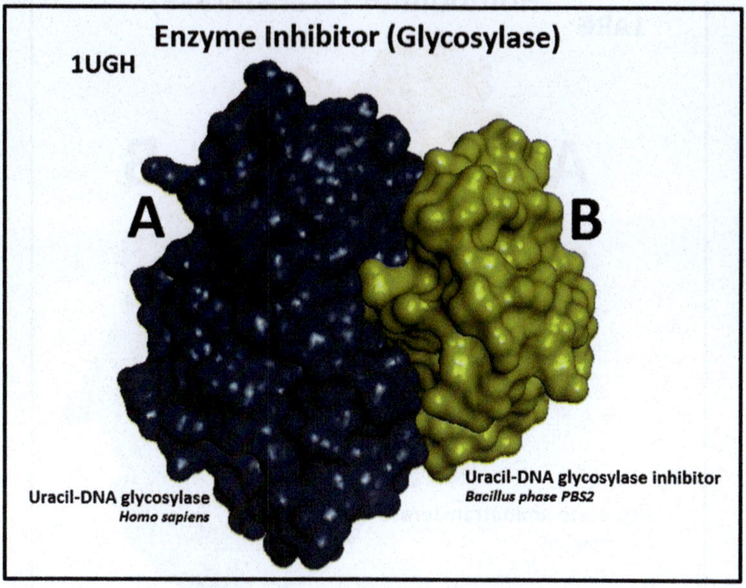

Fig. 2.5 Enzyme inhibitor (glycosylase) with PDB ID: 1UGH having chain A and chain B colored blue and yellow, respectively, is shown using Discovery Studio™

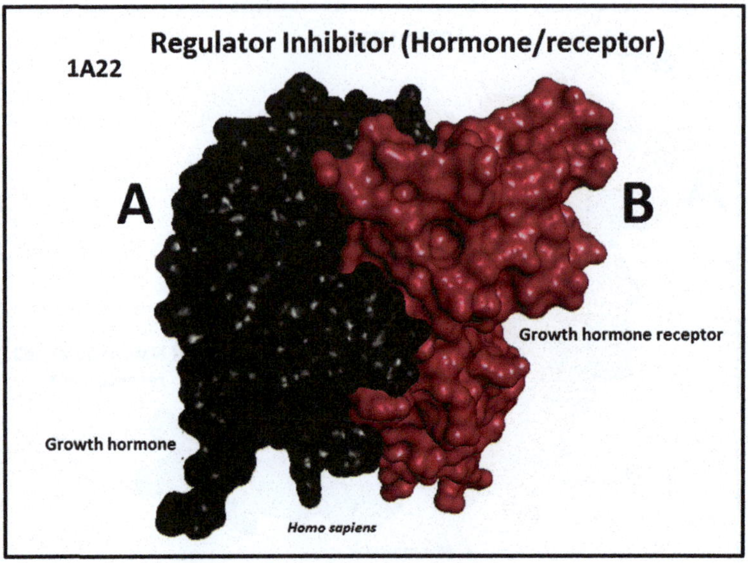

Fig. 2.6 Regulator inhibitor (hormone/receptor) with PDB ID: 1A22 having chain A and chain B colored black and magenta, respectively, is shown using Discovery Studio™

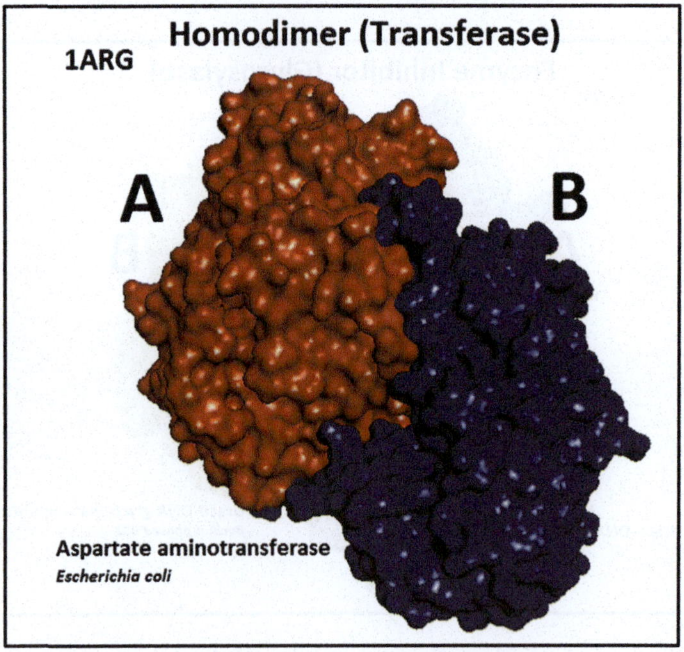

Fig. 2.7 Homodimer (transferase) complex with PDB ID: 1ARG having chain A and chain B colored dark orange and dark blue, respectively, is shown using Discovery Studio™

Homodimeric protein (1KSO) pairwise alignment

```
1KSO_chainA     1 MARPLEQAVAAIVCTFQEYAGRCGDKYKLCQAELKELLQKELATWTPTEF     50
                  ||||||||||||||||||||||||||||||||||||||||||||||||||
1KSO_chainB     1 MARPLEQAVAAIVCTFQEYAGRCGDKYKLCQAELKELLQKELATWTPTEF     50

1KSO_chainA    51 RECDYNKFMSVLDTNKDCEVDFVEYVRSLACLCLYCHEYFKDCPSEPPCS    100
                  ||||||||||||||||||||||||||||||||||||||||||||||||||
1KSO_chainB    51 RECDYNKFMSVLDTNKDCEVDFVEYVRSLACLCLYCHEYFKDCPSEPPCS    100

1KSO_chainA   101 Q      101          Identity - 100%
                  |
1KSO_chainB   101 Q      101          Similarity - 100%
```

Fig. 2.8 Pair-wise sequence alignment of a homodimeric protein (1KSO) is shown using the EMBOSS needle global alignment tool [29]

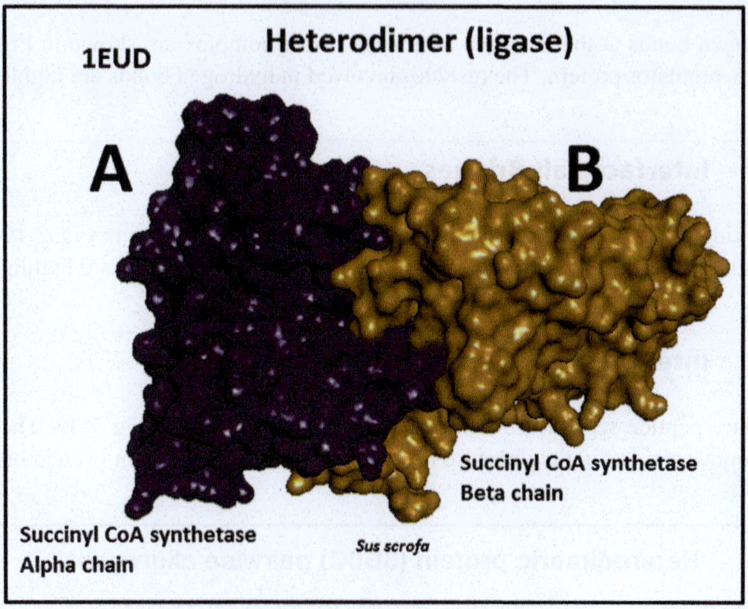

Fig. 2.9 Heterodimer (ligase) complex with PDB ID: 1EUD having chain A (violet) and chain B (light brown) is shown using Discovery Studio™

through the formation of a stable interface. Therefore, it is important to understand the molecular principles of protein-protein binding in relation to its function. The binding happens between identical subunits forming homodimers (Fig. 2.7) and between non-identical subunits forming heterodimers (Fig. 2.9).

2.2 Homodimer Protein Complex

A homodimer protein complex is made up of two (dimer) identical subunits as shown in Fig. 2.7. The sequence identity between these two subunits is 100% as shown in Fig. 2.8 and the subunits are homologous.

2.3 Heterodimer Protein Complex

A heterodimer protein complex is made up of two (dimer) non-identical subunits as shown in Fig. 2.9. The sequence identity between these two subunits is weak as shown in Fig. 2.10 and the subunits are non-homologous.

2.4 Interface Hydrogen Bonds

Hydrogen bonds at the interface of a heterodimer complex are shown in Fig. 2.11 using a regulator protein. The residues involved in hydrogen bonds are highlighted.

2.5 Interface Salt Bridges

Salt bridges at the interface of a heterodimer complex are shown in Fig. 2.12 using a regulator protein. The residues involved in salt bridge formation are highlighted.

2.6 Interface Peptide Segment

Interface peptide segments are illustrated using an example in Fig. 2.13. These are continuous regions in the interface that are associated with protein-protein binding.

Heterodimeric protein (3B0C) pairwise alignment

```
3B0C_chainT        1 GSTREPEIASSLIKQIFSHYVKTPVTRDAYKIVEKCSERYFKQISSDLEA     50

3B0C_chainW        1 --------------------------------------------------      0

3B0C_chainT       51 YSQHAGRKTVEMADVELLMR------RQGLVTDKMP-------LHVLVE-     86
                          ||:||....:..:::      |....||.:.       ||.|.|
3B0C_chainW        1 -----GRRTVPRGTLRKIIKKHKPHLRLAANTDLLVHLSFLLFLHRLAEE     45

3B0C_chainT       87 -RHLPLEYR-KLLIP-IAVSGNKVIPCK---    111   Identity - 14.5%
                      |....|.: |::.| ..::..|||..|
3B0C_chainW       46 ARTNAFENKSKIIKPEHTIAAAKVILKKSRG     76   Similarity - 22.9%
```

Fig. 2.10 Pair-wise sequence alignment of heterodimeric protein (3B0C) is shown using the EMBOSS needle global alignment tool [29]

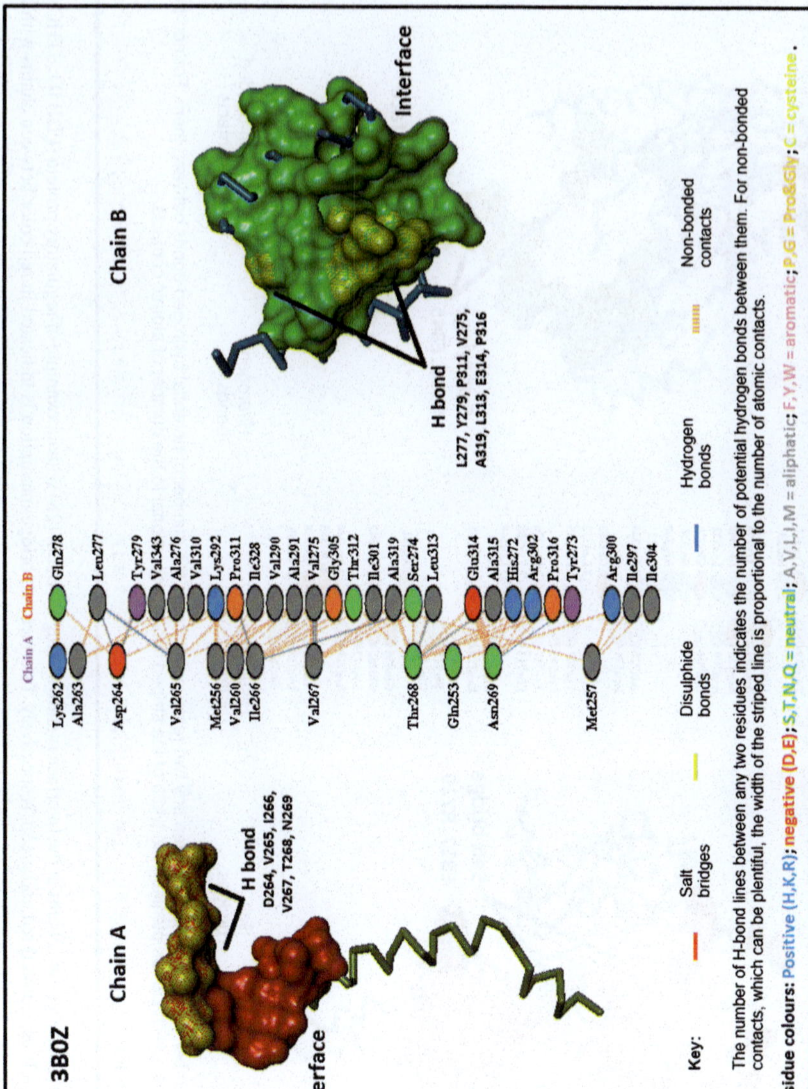

Fig. 2.11 Interface interaction between two chains is shown using PDBsum [28] using an example of regulatory protein (PDB ID: 3B0Z). Interface residues of chain A and chain B are colored red and green, respectively. Interface hydrogen bonds are indicated in blue color between chains with labeled amino acid residues

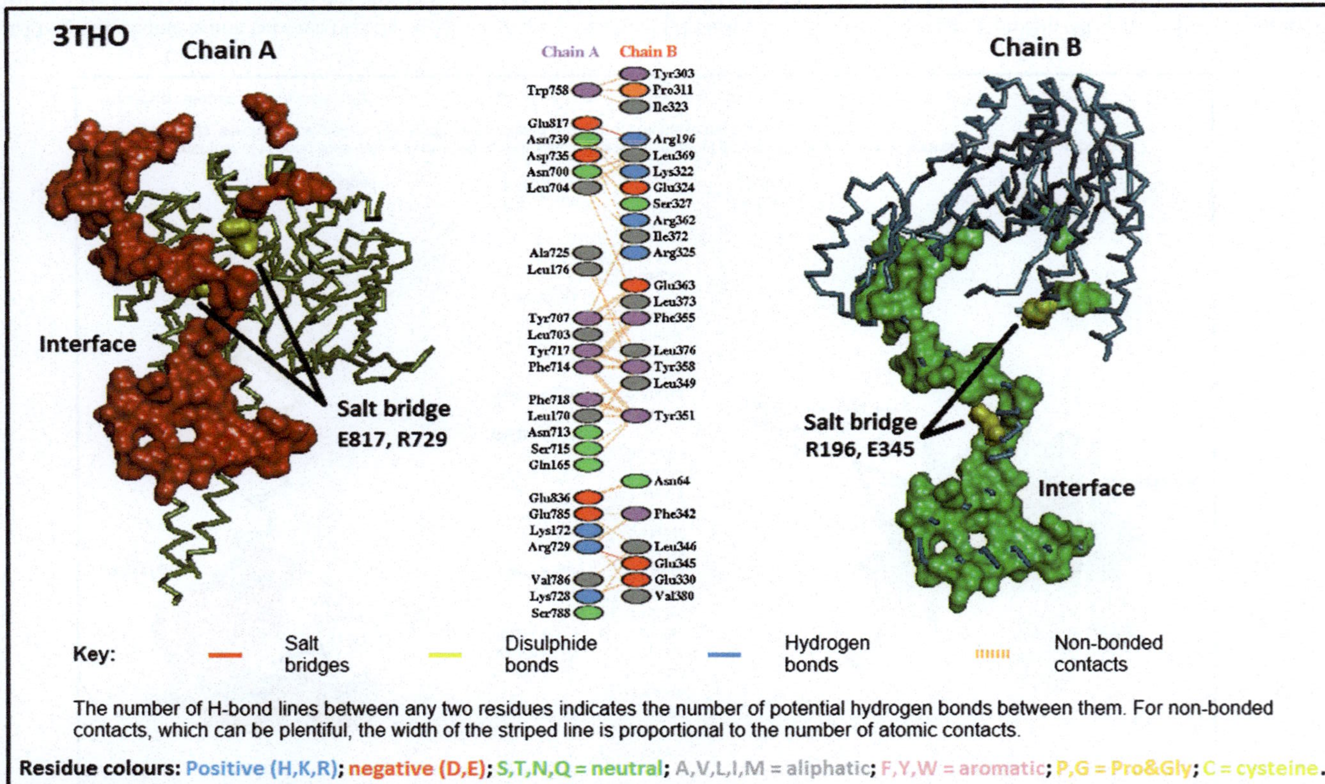

Fig. 2.12 Interaction between two chains at the interface is shown using PDBsum [28] with an example of regulatory protein (PDB ID: 3THO). Interface residues of chain A and chain B are colored red and green, respectively. Interface salt bridge interaction is indicated in red color between chains with labeled residues

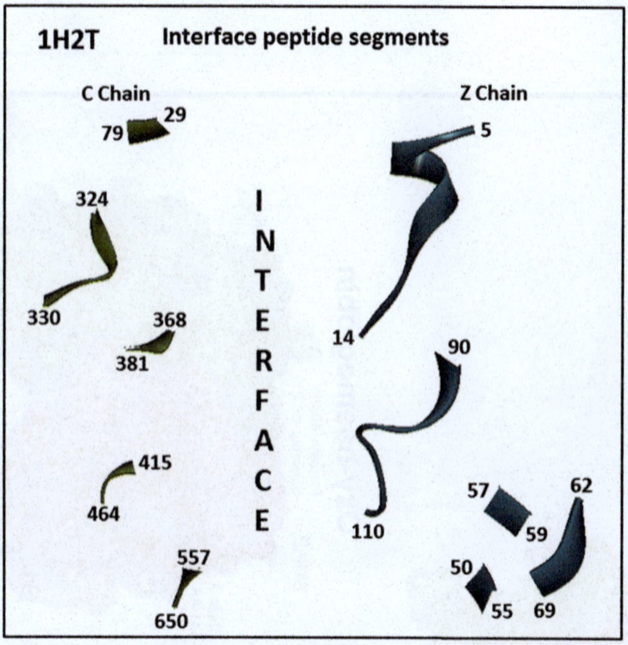

Fig. 2.13 Interface peptide segments (PDB ID: 1H2T) are shown using Discovery Studio™

2.7 Principles of Protein-Protein Binding

The broadly known features of protein-protein binding are schematically illustrated in Fig. 2.14 as documented by several groups (Table 2.1 and Table 2.2) and authors (Table 2.1) with dataset size (Table 2.1), nature of data (Table 2.3), and corresponding inference (Tables 2.4 and 2.5). A number of groups have used several datasets to draw conclusion on protein-protein interaction as summarized in Table 2.4. The classical work (Fig. 2.15) by Chothia and Janin [1] shows the presence of hydrophobicity at the interface with closely packed hydrophobic residues [2]. A number of observations have been documented since then on the principles of protein-protein binding (Table 2.4).

Table 2.5 classifies the dataset into different groups such as (a) homodimers, (b) heterodimers, (c) combination of heterodimers and homodimers, and (d) others. Homodimers have large interfaces with clusters [13] of hydrophobic residues [14]. Heterodimer interfaces are described by closely packed hydrophobic residues [1, 2]. Further, it was clarified that interface hydrophobicity is more than surface but less than core by Jones and Thornton [5]. The presence of hydrophobic interface with charged groups [11] was subsequently described. Interfaces are known to be stabilized by hydrogen bonds and salt bridges [7]. The amount of hydrogen bonds (Fig. 2.16) increases with interface size in both homo- and heterodimers [18]. The heterodimer interface is additionally characterized by several interface parameters [21], recognition sites [12], conserved residues [23], and peptide segments [20]

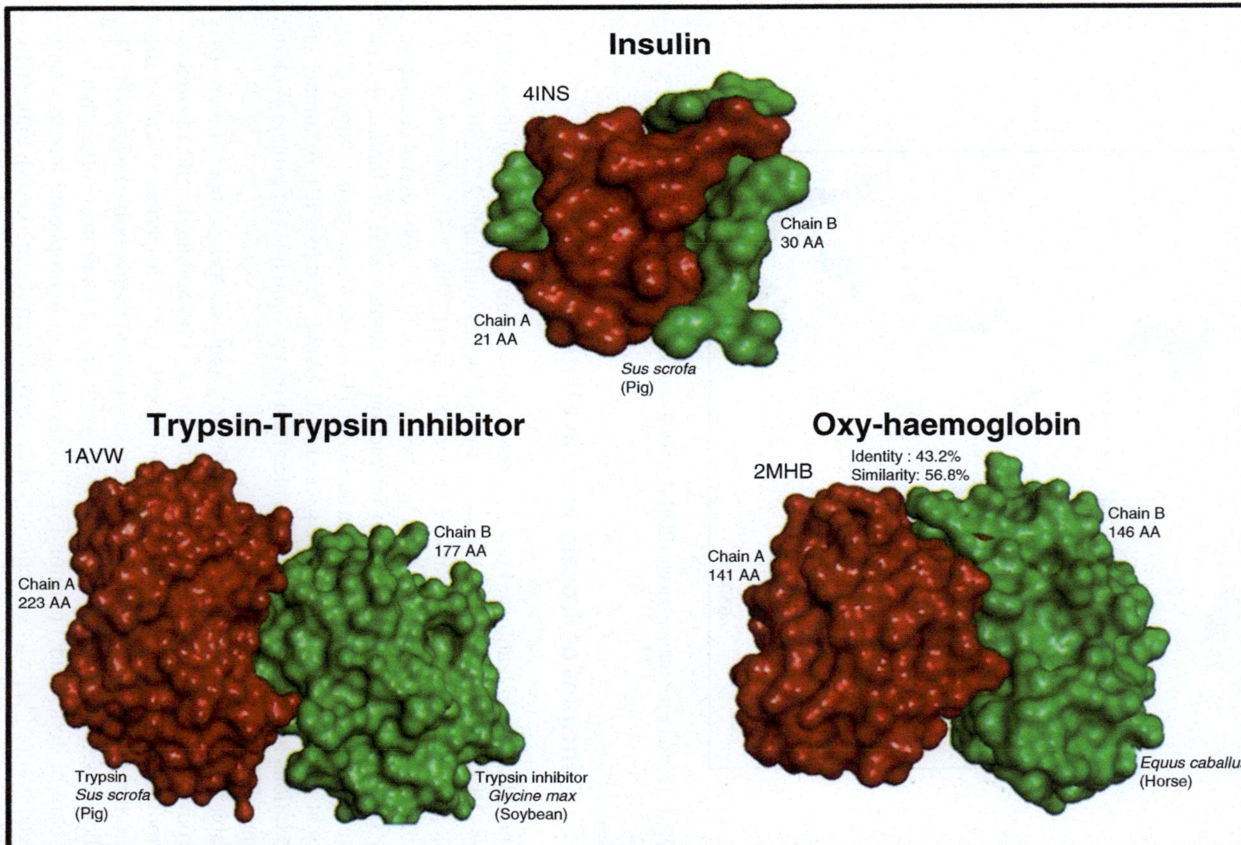

Fig. 2.14 A schematic understanding of an interface using known binding principles as reported by different groups (based on a rough classification) is shown. Please refer to Table 2.4 for further information on this illustration

Table 2.1 Protein-protein complexes studied by several authors from different groups with dataset size

S. no	Reference	Year	Country	Group	Data size
1	Chothia and Janin	1975	France	J. Janin	3
2	Chothia et al.	1976	France	J. Janin	2
3	Miller et al.	1987	UK	J. Janin	23
4	Janin and Chothia	1990	UK	J. Janin	20
5	Jones and Thornton	1995	UK	JM. Thornton	32
6	Jones and Thornton	1996	England	JM. Thornton	59
7	Xu et al.	1997	USA	R. Nussinov	319
8	Tsai et al.	1997	USA	R. Nussinov	419
9	Dasgupta et al.	1997	USA	JA. Bell	281
10	Linzaad and Argos	1997	Germany	P. Argos	218
11	Lo Conte et al.	1999	France	J. Janin	75
12	Chakrabarti and Janin	2002	France	J. Janin	70
13	Brinda et al.	2002	India	S. Vishveshwara	20
14	Bahadur et al.	2003	France	J. Janin	122
15	Nooren and Thornton	2003	UK	JM. Thornton	39
16	Caffrey et al.	2004	USA	ES. Huang	64
17	Bahadur et al.	2004	France	J. Janin	192
18	Zhanhua et al.	2005	Singapore	P. Kangueane	65
19	Zhanhua et al.	2005	Singapore	P. Kangueane	326
20	Pal et al.	2007	France	J. Janin	204
21	Reynolds et al.	2009	UK	S. Jones	754
22	Gromiha et al.	2009	Japan	K. Fukui	153
23	Guharoy and Chakrabarti	2010	India	P. Chakrabarti	326
24	Sowmya et al.	2011	India	P. Kangueane	192
25	Chen et al.	2013	USA	L. Regan	113
26	Sowmya et al.	2015	Australia	S. Ranganathan	278
27	Sowmya et al.	2015	Australia	S. Ranganathan	278

(Fig. 2.13). It is further reported that charged and aromatic residues initiate binding while hydrophobic residues assist in stability [22]. It is known that interface area and binding affinity are related [25]. The presence of interface polar residues more than surface and core in some complexes through residue level analysis is interesting [24, 26]. The relationship between molecular function and interface features is also established [27]. It should be noted that interface size relates to interface area in protein-protein complexes as shown in Fig. 2.17. Thus, the protein-protein interface is characterized by closely packed hydrophobic features with stabilizing hydrogen bonds and salt bridges having distributed electrostatic properties creating suitable recognition sites in conserved clustered groups is documented.

Table 2.2 Group-wise contribution to the understanding of protein-protein binding

Citation	Year	Country	Group	Data size
Chothia and Janin	1975	France	J. Janin	3
Chothia et al.	1976	France	J. Janin	2
Miller et al.	1987	UK	J. Janin	23
Janin and Chothia	1990	UK	J. Janin	20
Lo Conte et al.	1999	France	J. Janin	75
Chakrabarti and Janin	2002	France	J. Janin	70
Bahadur et al.	2003	France	J. Janin	122
Bahadur et al.	2004	France	J. Janin	192
Pal et al.	2007	France	J. Janin	204
Guharoy and Chakrabarti	2010	India	P. Chakrabarti	326
Jones and Thornton	1995	UK	JM. Thornton	32
Jones and Thornton	1996	England	JM. Thornton	59
Nooren and Thornton	2003	UK	JM. Thornton	39
Reynolds et al.	2009	UK	S. Jones	754
Xu et al.	1997	USA	R. Nussinov	319
Tsai et al.	1997	USA	R. Nussinov	419
Zhanhua et al.	2005	Singapore	P. Kangueane	65
Zhanhua et al.	2005	Singapore	P. Kangueane	326
Sowmya et al.	2011	India	P. Kangueane	192
Sowmya et al.	2015	Australia	S. Ranganathan	278
Sowmya et al.	2015	Australia	S. Ranganathan	278
Brinda et al.	2002	India	S. Vishveshwara	20
Dasgupta et al.	1997	USA	JA. Bell	281
Caffrey et al.	2004	USA	ES. Huang	64
Gromiha et al.	2009	Japan	K. Fukui	153
Chen et al.	2013	USA	L. Regan	113
Linzaad and Argos	1997	Germany	P. Argos	218

Table 2.3 A detailed description of protein-protein complexes studied by several authors from different groups

S. no	Citation	Data size	Data types			References
			Homodimer	Heterodimer	Others	
1	Chothia and Janin	3	–	3	Insulin dimer, trypsin-PTI, α/β oxyhemoglobin	[1]
2	Chothia et al.	2	–	2	Horse methemoglobin, human hemoglobin	[2]
3	Miller et al.	23	–	–	11 dimers, 9 tetramers, 2 hexamers, 1 octamer	[3]
4	Janin and Chothia	20	–	–	15 protease inhibitors, 4 antigen-antibody complexes	[4]
5	Jones and Thornton	32	–	32	–	[5]

(continued)

Table 2.3 (continued)

S. no	Citation	Data size	Data types Homodimer	Heterodimer	Others	References
6	Jones and Thornton	59	–	59	–	[6]
7	Xu et al.	319	–	–	–	[7]
8	Tsai et al.	419	–	–	362 non-redundant protein-protein interfaces, 57 symmetry-related oligomeric interfaces	[8]
9	Dasgupta et al.	281	–	–	58 oligomeric proteins, 223 protein crystal structures	[9]
10	Linzaad and Argos	218	–	–	59 protein complexes and 159 polypeptide chains	[10]
11	Lo Conte et al.	75	–	–	24 protease inhibitors, 19 antigen- antibody, 32 others (9 enzyme inhibitors, 11 signal transduction)	[11]
12	Chakrabarti and Janin	70	–	70	–	[12]
13	Brinda et al.	20	20	–	–	[13]
14	Bahadur et al.	122	122	–	–	[14]
15	Nooren and Thornton	39	–	39	–	[15]
16	Caffrey et al.	64	42	12	+10 transient complexes	[16]
17	Bahadur et al.	192	122	70	–	[17]
18	Zhanhua et al.	65	–	65	–	[18]
19	Zhanhua et al.	326	170	156	–	[19]
20	Pal et al.	204		204	–	[20]
21	Reynolds et al.	754	534	220	–	[21]
22	Gromiha et al.	153	–	153	–	[22]
23	Guharoy and Chakrabarti	326	122	204	–	[23]
24	Sowmya et al.	192	–	192	–	[24]
25	Chen et al.	113	–	113	–	[25]
26	Sowmya et al.	278	–	278	–	[26]
27	Sowmya et al.	278	–	278	–	[27]

Table 2.4 Summary of inference made on protein-protein binding in chronological order

S. No	Citation	Year	Group	Data size	Inference interface
1	Chothia and Janin	1975	J. Janin	3	Hydrophobicity stabilizes
2	Chothia et al.	1976	J. Janin	2	Closely packed hydrophobic residues
3	Miller et al.	1987	J. Janin	23	Interface area to relative molecular mass
4	Janin and Chothia	1990	J. Janin	20	Hydrophobicity stabilizes
5	Jones and Thornton	1995	JM Thornton	32	Interface hydrophobicity more than surface but less than core
6	Jones and Thornton	1996	JM Thornton	59	Differentiates homodimers, heterodimers, Ag-Ab, enzyme inhibitors
7	Xu et al.	1997	R. Nussinov	319	Interface, hydrogen bonds and salt bridges
8	Tsai et al.	1997	R. Nussinov	419	Interface hydrophobicity
9	Dasgupta et al.	1997	JA. Bell	281	Hydrophobic interfaces favor aromatic and methionine over aliphatic residues in oligomer complexes
10	Linzaad and Argos	1997	P. Argos	218	Interface hydrophobic patches in prediction is demonstrated
11	Lo Conte et al.	1999	J. Janin	75	Hydrophobic interface with charged groups
12	Chakrabarti and Janin	2002	J. Janin	70	Interface recognition sites
13	Brinda et al.	2002	S. Vishveshwara	20	Amino acid clusters in homodimer interface
14	Bahadur et al.	2003	J. Janin	122	Hydrophobic residues in usually large homodimeric interfaces
15	Nooren and Thornton	2003	JM. Thornton	39	Conserved residues in homodimeric interfaces
16	Caffrey et al.	2004	ES. Huang	64	Interface residues are more conserved
17	Bahadur et al.	2004	J. Janin	192	Specific vs non-specific interfaces shown
18	Zhanhua et al.	2005	P. Kangueane	65	Homodimer vs heterodimer interface discussed
19	Zhanhua et al.	2005	P. Kangueane	326	Interface parameters that determine heterodimeric complexes
20	Pal et al.	2007	J. Janin	204	Peptide segments in protein interfaces
21	Reynolds et al.	2009	S. Jones	754	Interface analysis
22	Gromiha et al.	2009	K. Fukui	153	Charged and aromatic residues initiate binding, while hydrophobic residues assist stability

(continued)

Table 2.4 (continued)

S. No	Citation	Year	Group	Data size	Inference interface
23	Guharoy and Chakrabarti	2010	P. Chakrabarti	326	Conserved residue clusters at the interface
24	Sowmya et al.	2011	P. Kangueane	192	Interface polar residues more than surface and core
25	Chen et al.	2013	L. Regan	113	Interface area vs binding affinity
26	Sowmya et al.	2015	S. Ranganathan	278	Interface features vs biological function
27	Sowmya et al.	2015	S. Ranganathan	278	Residue level classification of interfaces

Table 2.5 Summary of inference made on protein-protein binding based on dataset nature

Citation	Year	Data types			Interface Inference
		Homodimer	Heterodimer	Others	
Brinda et al.	2002	20	–	–	Amino acid clusters
Bahadur et al.	2003	122	–	–	Hydrophobic residues in large interfaces
Chothia and Janin	1975	–	3	Insulin dimer, trypsin-PTI, α/β oxyhemoglobin	Hydrophobicity stabilizes
Chothia et al.	1976	–	2	Horse methemoglobin, human hemoglobin	Closely packed hydrophobic residues
Jones and Thornton	1995	–	32	–	Hydrophobicity more than surface but less than core
Jones and Thornton	1996	–	59	–	Homodimers, heterodimers, Ag-Ab, enzyme inhibitors
Chakrabarti and Janin	2002	–	70	–	Interface recognition sites
Nooren and Thornton	2003	–	39	–	Conserved residues
Zhanhua et al.	2005a	–	65	–	Interface parameters analysis
Pal et al.	2007	–	204	–	Peptide segments in protein interfaces
Gromiha et al.	2009	–	153	–	Charged and aromatic residues initiate binding while hydrophobic residues assists stability
Sowmya et al.	2011	–	192	–	Interface polar residues more than surface and core

(continued)

Table 2.5 (continued)

Citation	Year	Data types			Interface Inference
		Homodimer	Heterodimer	Others	
Chen et al.	2013	–	113	–	Interface area vs binding affinity
Sowmya et al.	2015	–	278	–	Interface features vs biological function
Sowmya et al.	2015	–	278	–	Residue level classification of interfaces
Bahadur et al.	2004	122	70	–	Specific vs non-specific interfaces
Zhanhua et al.	2005b	170	156	–	Hydrogen bonds
Reynolds et al.	2009	534	220	–	Interface analysis
Guharoy and Chakrabarti	2010	122	204	–	Conserved residue clusters at the interface
Caffrey et al.	2004	42	12	+10 transient complexes	Interface residues are more conserved
Xu et al.	1997	–	–	–	Interface, hydrogen bonds and salt bridges
Tsai et al.	1997	–	–	362 non-redundant protein-protein interfaces, 57 symmetry-related oligomeric interfaces	Interface hydrophobicity
Dasgupta et al.	1997	–	–	58 oligomeric proteins, 223 protein crystal structures	Hydrophobic interfaces favor aromatic and methionine over aliphatic residues in oligomer complexes
Linzaad and Argos	1997	–	–	59 protein complexes and 159 polypeptide chains	Interface hydrophobic patches in prediction is demonstrated
Lo Conte et al.	1999	–	–	24 protease inhibitors, 19 antigen- antibody, 32 others (9 enzyme inhibitors, 11 signal transduction)	Hydrophobic interface with charged groups
Miller et al.	1987	–	–	11 dimers, 9 tetramers, 2 hexamers, 1 octamer	Interface area is related to relative molecular mass
Janin and Chothia	1990	–	–	15 protease inhibitors, 4 antigen-antibody complexes	Hydrophobicity stabilizes

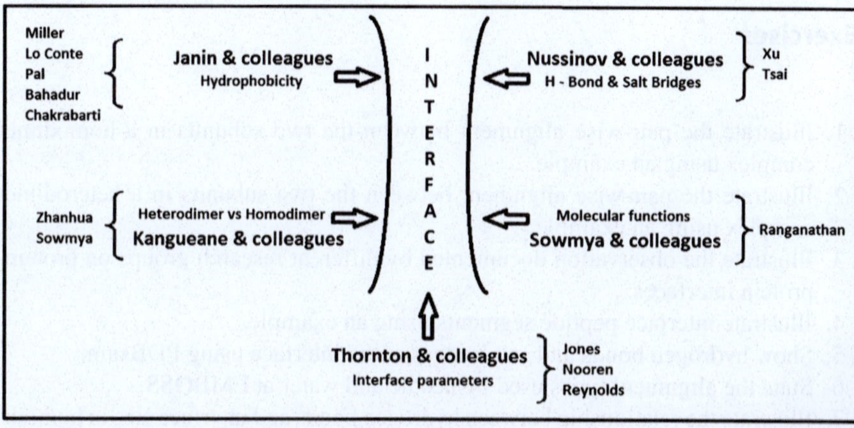

Fig. 2.15 Insulin (PDB ID: 4INS), trypsin-trypsin inhibitor (PDB ID: 1AVW), and oxyhemoglobin (PDB ID: 2MHB) are the three protein complexes studied by Chothia and Janin [1]. Chains A and B are shown in green and red color, respectively, using Discovery Studio™

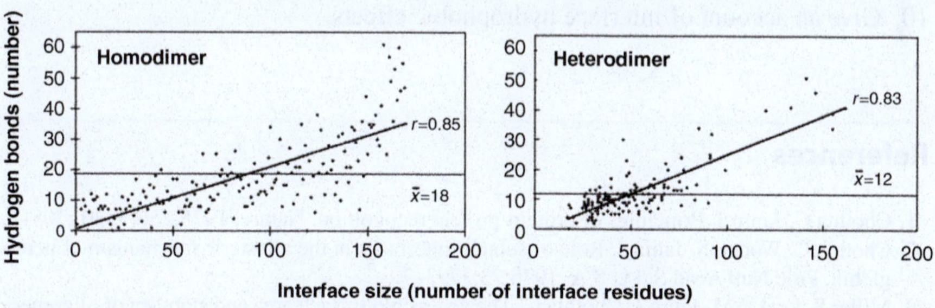

Fig. 2.16 Relationship between interface size (number of interface residues) and hydrogen bonds in homodimer and heterodimer is shown. This is adapted from Zhanhua et al. [18] under the open access creative commons attribution license

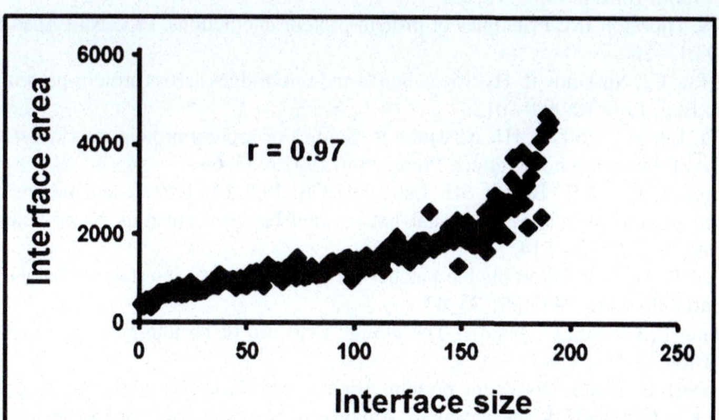

Fig. 2.17 Relationship between interface size (number of interface residues) and interface areas is shown. This is adapted from Sowmya et al. [24] under the open access creative commons attribution license

Exercises

1. Illustrate the pair-wise alignment between the two subunits in a homodimer complex using an example.
2. Illustrate the pair-wise alignment between the two subunits in a heterodimer complex using an example.
3. Illustrate the observation documented by different research groups on protein-protein interfaces.
4. Illustrate interface peptide segments using an example.
5. Show hydrogen bonds and salt bridges at the interface using PDBsum.
6. State the alignment types used by needle and water at EMBOSS.
7. Illustrate the relationship between hydrogen bonds and interface size in homodimers and heterodimers and describe in detail.
8. Illustrate the molecular functions of protein complexes using appropriate examples.
9. Differentiate between homodimers and heterodimers.
10. Give an account of interface hydrophobic effects.

References

1. Chothia C, Janin J. Principles of protein-protein recognition. Nature. 1975;256(5520):705–8.
2. Chothia C, Wodak S, Janin J. Role of subunit interfaces in the allosteric mechanism of hemoglobin. Proc Natl Acad Sci U S A. 1976;73:3793–7.
3. Miller S, Lesk AM, Janin J, Chothia C. The accessible surface area and stability of oligomeric proteins. Nature. 1987;328(6133):834–6.
4. Janin J, Chothia C. The structure of protein-protein recognition sites. J Biol Chem. 1990;265(27):16027–30.
5. Jones S, Thornton JM. Protein-protein interactions: a review of protein dimer structures. Prog Biophys Mol Biol. 1995;63:31–65.
6. Jones S, Thornton JM. Principles of protein-protein interactions. Proc Natl Acad Sci U S A. 1996;93:13–20.
7. Xu D, Tsai CJ, Nussinov R. Hydrogen bonds and salt bridges across protein-protein interfaces. Protein Eng. 1997;10:999–1012.
8. Tsai CJ, Lin SL, Wolfson HJ, Nussinov R. Studies of protein-protein interfaces: a statistical analysis of the hydrophobic effect. Protein Sci. 1997;6:53–64.
9. Dasgupta S, Iyer GH, Bryant SH, Lawrence CE, Bell JA. Extend and nature of contacts between protein molecules in crystal lattices and between subunits of protein oligomers. Proteins. 1997;28:494–514.
10. Lijnzaad P, Argos P. Hydrophobic patches on protein subunit interfaces: characteristics and prediction. Proteins. 1997;28:333–43.
11. Lo Conte L, Chothia C, Janin J. The atomic structure of protein-protein recognition sites. J Mol Biol. 1999;285:2177–98.
12. Chakrabarti P, Janin J. Dissecting protein-protein recognition sites. Proteins. 2002;47:334–43.
13. Brinda KV, Kannan N, Vishveshwara S. Analysis of homodimeric protein interfaces by graph-spectral methods. Protein Eng. 2002;15:265–77.

14. Bahadur RP, Chakrabarti P, Rodier F, Janin J. Dissecting subunit interfaces in homodimeric proteins. Proteins. 2003;53:708–19.
15. Nooren IM, Thornton JM. Structural characterization and functional significance of transient protein-protein interactions. J Mol Biol. 2003;325:991–1018.
16. Caffrey DR, Somaroo S, Hughes JD, Mintseris J, Huang ES. Are protein-protein interfaces more conserved in sequence than the rest of the protein surface? Protein Sci. 2004;13:190–202.
17. Bahadur RP, Chakrabarti P, Rodier F, Janin JA. Dissection of specific and non-specific protein-protein interfaces. J Mol Biol. 2004;336:943–55.
18. Zhanhua C, Gan JG, Lei L, Mathura VS, Sakharkar MK, Kangueane P. Protein subunit interfaces: heterodimers versus homodimers. Bioinformation. 2005b;1:28–39.
19. Zhanhua C, Gan JG, Lei L, Mathura VS, Sakharkar MK, Kangueane P. Identification of critical heterodimer protein interface parameters by multi-dimensional scaling in euclidian space. Front Biosci. 2005a;10:844–52.
20. Pal A, Chakrabarti P, Bahadur R, Rodier F, Janin J. Peptide segments in protein-protein interfaces. J Biosci. 2007;32:101–11.
21. Reynolds C, Damerell D, Jones S. ProtorP: a protein-protein interaction analysis server. Bioinformatics. 2009;25:413–4.
22. Gromiha MM, Yokota K, Fukui K. Energy based approach for understanding the recognition mechanism in protein-protein complexes. Mol BioSyst. 2009;5:1779–86.
23. Guharoy M, Chakrabarti P. Conserved residue clusters at protein-protein interfaces and their use in binding site identification. BMC Bioinform. 2010;11:286.
24. Sowmya G, Anita S, Kagueane P. Insights from the structural analysis of protein heterodimer interfaces. Bioinformation. 2011;6(4):137–43.
25. Chen J, Sawyer N, Regan L. Protein-protein interactions: general trends in the relationship between binding affinity and interfacial buried surface area. Protein Sci. 2013;22:510–5.
26. Sowmya G, Ranganathan S. Discrete structural features among interface residue-level classes. BMC Bioinform. 2015;16(Suppl 18):S8.
27. Sowmya G, Breen EJ, Ranganathan S. Linking structural features of protein complexes and biological function. Protein Sci. 2015;24:1486–94.
28. Laskowski RA. PDBsum new things. Nucleic Acids Res. 2009; 37(Database):D355–9.
29. Needleman SB, Wunsch CDA. General method applicable to the search for similarities in the amino acid sequence of two proteins. J Mol Biol. 1970;48(3):443–53. Available at http://www.ebi.ac.uk/Tools/psa/emboss-needle/

3 Low-Resolution Protein Complexes

3.1 Protein Complexes

Protein complexes play an important role in biological function. Different forms of complexes formed by several subunits of proteins are available in nature. Advancement in biotechnology, downstream processing, protein purification, and X-ray crystallization techniques have led to the solution of thousands of protein complex structures that are made available at the protein data bank (PDB) [1]. These high-resolution (<3 Å) complexes provide valuable molecular insights to their biological role. However, there are also few thousands of low-resolution (> = 3 Å) complexes at the PDB. We gleaned a non-redundant dataset of low-resolution complexes as shown in Fig. 3.1 for hetero complexes and Fig. 3.2 for homo complexes. Figure 3.1 shows the procedure for the creation of 592 hetero complexes. The distribution of hetero complexes based on the number of subunits in the complex is shown in Fig. 3.3. Similarly, Fig. 3.2 shows the procedure for the creation of 1463 homo complexes. The distribution of homo complexes based on the number of subunits in complex formation is also shown in Fig. 3.3.

3.2 Protein Hetero Complexes

There are about 600 low-resolution non-redundant hetero complexes (Table 3.1) at the protein data bank (PDB) as shown in Fig. 3.1. The distribution of these complexes based on the number of subunits is shown in Fig. 3.3. Dimer protein complexes (AB) formed by 2 subunits are the most represented in this group. It should also be noted that A2B2 complexes are second largest in the dataset. Thus, the frequency of occurrence of AB and A2B2 complexes is evident. The presence of A2B and ABC trimer protein complexes is intriguing.

P. Kangueane, C. Nilofer, *Protein-Protein and Domain-Domain Interactions*,
https://doi.org/10.1007/978-981-10-7347-2_3

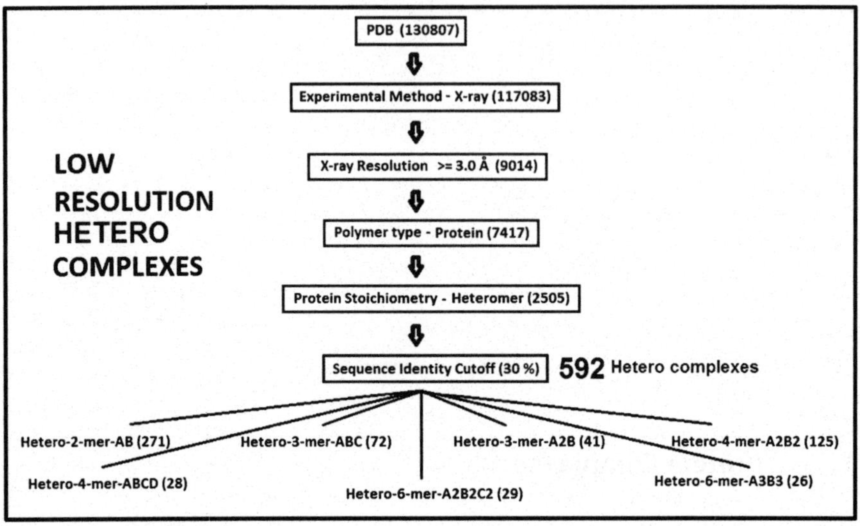

Fig. 3.1 A flowchart for the creation of a non-redundant low-resolution (> = 3 Å) hetero protein complexes from protein data bank (PDB) is made available at www.rcsb.org/pdb [1]

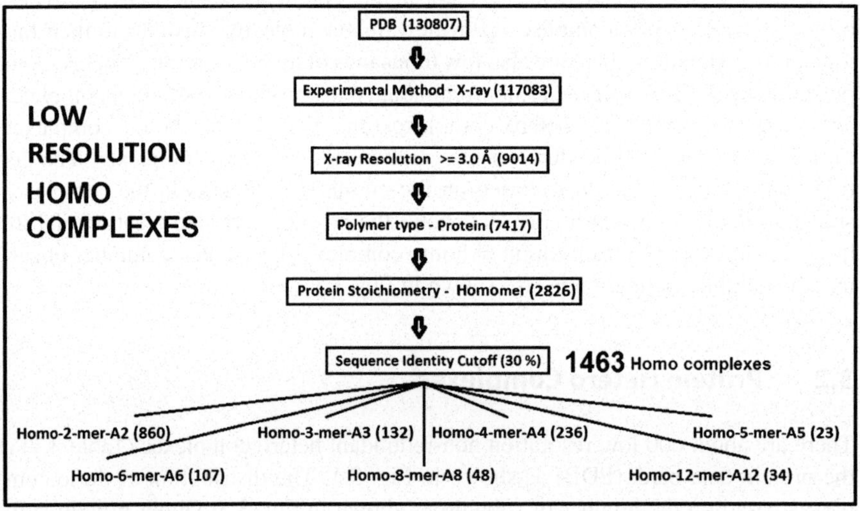

Fig. 3.2 A flowchart for the creation of non-redundant low-resolution (> = 3 Å) homo protein complexes from protein data bank (PDB) is made available at www.rcsb.org/pdb/ [1]

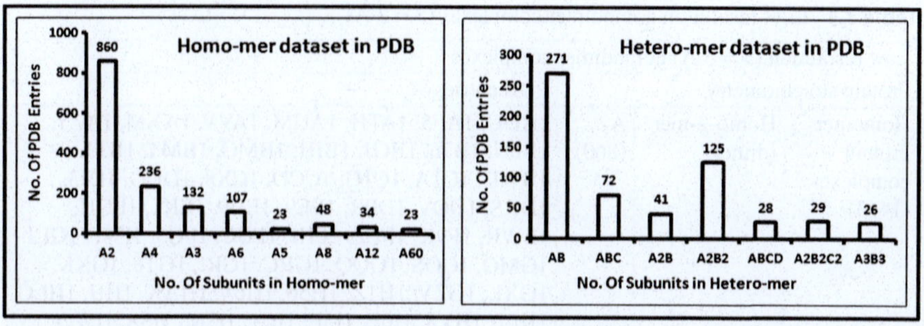

Fig. 3.3 Distribution of a non-redundant set of low-resolution (> = 3 Å) homo and hetero complexes at the protein data bank (PDB). Dimers are most represented for both homo and hetero protein complexes. It should be noted that ABC and A2B hetero trimer protein complexes are also available

3.2.1 AB Dimer Protein Complex

Figure 3.4 shows the structure of a heterodimer (2 subunits) protein complex (AB) of a transcription regulation complex with PDB ID: 1BH8 solved at 3.04 Å. This is a AB heterodimer complex made of 2 different non-homologous subunits. There are about 271 such AB hetero complexes in the dataset.

3.2.2 A2B Trimer Protein Complex

Figure 3.5 shows the structure of a hetero trimer (3 subunits) protein complex (A2B) of a signaling protein with PDB ID: 1NVX solved at 3.2 Å. This is a A2B hetero trimer complex made of 2 identical subunits and 1 subunit is non-identical. There are about 41 such A2B complexes in the dataset.

3.2.3 ABC Trimer Protein Complex

Figure 3.6 shows the structure of a hetero trimer (3 subunits) transport protein complex (ABC) with PDB ID: 1RH5 solved at 3.2 Å. It should be noted that all the three chains are non-identical in this complex. There are about 72 such ABC trimer complexes in the dataset.

3.2.4 ABCD Tetramer Protein Complex

Figure 3.7 shows the structure of a hetero tetramer (4 subunits) protein complex of a ligase enzyme (ABCD) with PDB ID: 1Z5S solved at resolution at 3 Å. It should be noted that all the 4 subunits are non-identical in this complex. There are about 28 such ABCD hetero complexes in the dataset.

Table 3.1 Low-resolution homodimer complexes (> = 3 Å)

Low resolution (> = 3 Å) homodimer complexes			
Protein stoichiometry			PDB entries
Homomer protein complexes (1463)	Homo-2-mer (dimer)	A2 (860)	1ADU, 1AF5, 1ATH, 1AUM, 1AVV, 1AXM, 1AZY, 1B6U, 1B76, 1BGJ, 1BIH, 1BMO, 1BMT, 1BTJ, 1BWD, 1C1A, 1C1G, 1CCD, 1CX8, 1DDO, 1DJ3, 1DKS, 1DOV, 1DPR, 1EE9, 1EH9, 1EK1, 1EQU, 1EWR, 1F9K, 1FEZ, 1FHI, 1FOC, 1FQ3, 1FVF, 1GLJ, 1GMO, 1GOS, 1GQQ, 1GRC, 1GRI, 1GT8, 1GXK, 1GXL, 1GZV, 1H1Z, 1H56, 1H6V, 1HAK, 1HIJ, 1HLG, 1HLP, 1HV8, 1I0E, 1I5E, 1IC1, 1ISM, 1ISS, 1IWO, 1J0K, 1J7I, 1JIJ, 1JJK, 1JL9, 1JQ7, 1JXA, 1JYF, 1K23, 1KA4, 1KAN, 1KAQ, 1KO6, 1KOA, 1LL0, 1LM7, 1LRP, 1M6E, 1M74, 1MAW, 1MC8, 1MIW, 1MO2,
			1MXC, 1N21, 1N2L, 1NJ8, 1NYH, 1NYQ, 1OK9, 1ORD, 1OSI, 1OU5, 1OVU, 1P75, 1PDG, 1PEO, 1PFC, 1PGQ, 1PYP, 1QB3, 1QC9, 1QOL, 1QX7, 1R30, 1RFB, 1RFQ, 1RGB, 1RLV, 1RPT, 1S28, 1S60, 1S9C, 1SJP, 1SMQ, 1SMS, 1SXI, 1T62, 1TAT, 1TDA, 1TUY, 1TVE, 1TX9, 1U9O, 1UAZ, 1UD0, 1UN9, 1UX4, 1UZG, 1V1F, 1VCQ, 1VG2, 1VGZ, 1VZ7, 1VZV, 1W1M, 1W3Z, 1WA9, 1WAT, 1WC4, 1X03, 1X24, 1XB4, 1XCP, 1XDM, 1XDO, 1XV8, 1Y11, 1Y1U, 1Y5R, 1Y6E, 1YCF, 1YFY, 1YPO, 1YW6, 1Z1X, 1Z8L, 1ZAK, 1ZOO, 1ZP6, 1ZTV, 1ZVU, 1ZY1, 1ZYC, 2A30, 2A3L, 2A62, 2A87, 2AFV, 2AKQ, 2AN6, 2AW6, 2AYI, 2AYU, 2B48, 2B7L, 2B92, 2BAP, 2BHV, 2C6X, 2CBF, 2CKP, 2CLP, 2CVT, 2CZI, 2D4Z, 2D7H, 2DDW, 2DEO, 2DH6, 2DQR, 2DVY, 2E7I, 2E7S, 2F3X, 2FNI, 2FT3, 2FXT, 2G4C, 2G6T, 2GFP, 2GUJ, 2GW1, 2GXF, 2H5K, 2H8K, 2HCK, 2HYD, 2HZ6, 2I76, 2IA9, 2IES, 2IF7, 2IJD, 2IKC, 2IOP, 2IS3, 2IWH, 2J68, 2JA3, 2JAA, 2JAX, 2JGT, 2JKZ, 2LJR, 2NRP, 2O1T, 2O7P, 2O96, 2OGU, 2OKH, 2ON3, 2OOE, 2OTO, 2OWM, 2P2T, 2P7K, 2PAF, 2PDA, 2PEK, 2PR1, 2PV3, 2PX0, 2Q47, 2Q5H, 2QFI, 2QYG, 2R6F, 2R6M, 2RAK, 2RKO, 2UPJ, 2V5R, 2V5Y, 2VAQ, 2VF8, 2VXC, 2VZ9, 2WDW, 2WFD, 2WFN, 2WKG, 2WYO, 2X0S, 2X6I, 2X72, 2X8C, 2XCO, 2XG7, 2XKP, 2XLS, 2XR7, 2XXY, 2XZ0, 2Y92, 2YCB, 2YHN, 2YMB, 2YPZ, 2YQ3, 2YVX, 2YVZ, 2YWA, 2Z0I, 2Z2R, 2Z87, 2ZB5, 2ZGK, 2ZHH, 2ZJG, 2ZR3, 2ZRT, 2ZUD, 2ZVF, 2ZXH, 3A44, 3A6M, 3A7M, 3AP3, 3AQL, 3AQS, 3AVQ, 3B4R, 3B57, 3B5D, 3B6A, 3BAC, 3BG9, 3BOA, 3BR6, 3BXZ, 3C3E, 3C51, 3CAZ, 3CEJ, 3CRC, 3CSM, 3CWO, 3D45, 3DCA, 3DEV, 3DJW, 3DKW, 3DMK, 3DTJ, 3E0Y, 3E1K, 3E5Q, 3E9Y, 3EHF, 3EIU, 3EQ2, 3EYL, 3F8N, 3FF6, 3FNU, 3FX0, 3G05, 3G6B, 3G8E, 3GEH, 3GFQ, 3GMH, 3GSB, 3GUW, 3GYV, 3H0K, 3H2T, 3H44, 3HBB, 3HGQ, 3HJ5, 3HNE, 3HQO, 3HRK, 3HUD, 3HZR, 3I64, 3I6U, 3IBJ, 3IBP, 3ICX, 3ID2, 3IHK,

(continued)

Table 3.1 (continued)

Low resolution (> = 3 Å) homodimer complexes			
Protein stoichiometry			PDB entries
			3IKL, 3IL4, 3IQW, 3JQP, 3JQZ, 3JUV, 3K1P, 3K55, 3K6H, 3KBP, 3KG5, 3KNF, 3KP5, 3KRN, 3KS2, 3KZF, 3L1G, 3L1L, 3L2J, 3L3P, 3L8J, 3LC6, 3LJL, 3LN7, 3LXA, 3LXU, 3M4Q, 3MBO, 3MCD, 3MCP, 3MF0, 3MIV, 3N4Q, 3N4X, 3N50, 3N5U, 3N6B, 3N6V, 3NAS, 3NR7, 3NTJ, 3NYJ, 3NZP, 3O9U, 3OE6, 3OHN, 3OOC, 3ORG, 3OZA, 3P0U, 3P40, 3P5N, 3PI4, 3PJZ, 3PQ1, 3PRD, 3Q0X, 3Q2W, 3Q41, 3QFG, 3QJO, 3QK9, 3QKV, 3QLK, 3QMO, 3QNQ, 3QOE, 3QP5, 3QQ2, 3R6M, 3RB9, 3RBL, 3RFG, 3RG2, 3RH7, 3RJR, 3RYL, 3S0P, 3S4D, 3S4U, 3S6H, 3SC3, 3SE7, 3SEJ, 3SN9, 3SNA, 3SNH, 3SVA, 3SYL, 3T1I, 3T35, 3T5D, 3TAZ, 3TH1, 3TSY, 3TVI, 3TY9, 3U1N, 3U3T, 3UA3, 3UGT, 3UKM, 3UM0, 3UQH, 3UT7, 3UZB, 3V4L, 3V7S, 3VAZ, 3VEM, 3VH1, 3VOX, 3W03, 3WCN, 3WTF, 3WVO, 3WZH, 3X21, 3ZJR, 3ZRT, 4A2M, 4A5M, 4A7P, 4AH6, 4AKG, 4ALB, 4AP5, 4AQB, 4AQT, 4ASN, 4ATV, 4AV7, 4B09, 4B56, 4BBD, 4BDJ, 4BDT, 4BEJ, 4BEP, 4BIY, 4BMG, 4BTY, 4BW5, 4BWP, 4BXD, 4C51, 4C8U, 4CBV, 4CID, 4CJ7, 4CLV, 4CTD, 4CUE, 4CWE, 4CZ8, 4D1E, 4D28, 4D4J, 4D5S, 4D60, 4D87, 4D8U, 4DDQ, 4DTC, 4DUL, 4DVD, 4DZW, 4E0I, 4E0V, 4E3C, 4E85, 4E8Z, 4EA3, 4EGB, 4EIA,
			4ET0, 4F2H, 4F2Z, 4F35, 4F5D, 4F7R, 4FB8, 4FE4, 4FMF, 4FN0, 4FN9, 4FPB, 4FQU, 4FR2, 4FS9, 4FSC, 4FT2, 4G12, 4G3K, 4G85, 4GA5, 4GAV, 4GFA, 4GHS, 4GI2, 4GKW, 4GO4, 4GPO, 4GQX, 4GU7, 4GWN, 4GZ8, 4H54, 4H8S, 4HBO, 4HG0, 4HQR, 4HRH, 4I14, 4I5S, 4I9G, 4IGG, 4IIA, 4ILD, 4ILR, 4J2Q, 4J72, 4JDQ, 4JEN, 4JGM, 4JGV, 4JHJ, 4JLC, 4JM0, 4JUP, 4K03, 4KGO, 4KK1, 4KM3, 4KMH, 4KP4, 4KPN, 4KYA, 4KYN, 4L27, 4L6Y, 4L9J, 4L9M, 4LD4, 4LDB, 4LDS, 4LEI, 4LJ2, 4LMV, 4LPU, 4LU6, 4LZP, 4M8M, 4ML3, 4MVC, 4MWT, 4MX3, 4N1K, 4NDW, 4NFR, 4NIC, 4NK7, 4NT5, 4O1O, 4O6B, 4O6O, 4OAA, 4OFD, 4OH3, 4OIZ, 4OMT, 4OPH, 4OT9, 4P1O, 4P8Q, 4PBB, 4PK6, 4PL4, 4PMV, 4PN6, 4Q0C, 4Q2G, 4Q37, 4Q7C, 4Q85, 4QEB, 4QQV, 4R1I, 4R5G, 4RCK, 4RG7, 4RSJ, 4RT7, 4RUE, 4RY2, 4RYI, 4S0P, 4TKR, 4TN3, 4TPH, 4TWA, 4U5T, 4UBF, 4UBU, 4UD0, 4UF2, 4UOI, 4UP9, 4UQV, 4UZV, 4V1V, 4W6S, 4WHJ, 4WIB, 4WIJ, 4WIK, 4WIO, 4WIS, 4WP0, 4WPC, 4WRN, 4X0F, 4X2E, 4X5N, 4X8C, 4X8F, 4XA1, 4XAJ, 4XAU, 4XC6, 4XHA, 4XIV, 4XJR, 4XKJ, 4XRX, 4XWS, 4Y13, 4Y4R, 4Y7F, 4YCV, 4YDG, 4YF6, 4YFF, 4YFG, 4YH6, 4YIY, 4YNA, 4YO3, 4YRB, 4YVC, 4YZY, 4Z1I, 4Z25, 4Z2Y, 4Z5V, 4Z7F, 4ZDI, 4ZMX, 4ZS5, 4ZZH, 5A0Z, 5A9Y, 5AE4, 5AFR, 5AJV, 5AKP, 5B01, 5B0M, 5B1I, 5B37, 5BP4, 5BW7, 5BZ2, 5C5K, 5C76, 5CB0, 5CCA, 5CF2, 5CIM, 5CJ4, 5CKQ, 5CLR, 5CO7, 5CQR,

(continued)

Table 3.1 (continued)

Low resolution (> = 3 Å) homodimer complexes			
Protein stoichiometry			PDB entries
			5CRE, 5CSA, 5CSK, 5CV3, 5D0N, 5D1W, 5D7B, 5D8E, 5DBK, 5DG1, 5DGY, 5DL2, 5DPD, 5DQA, 5DZL, 5E0O, 5E3S, 5E8K, 5EG1, 5EN6, 5EPG, 5EQ6, 5ERF, 5F4U, 5F4Y, 5FIY, 5FJY, 5FM2, 5FPH, 5FSW, 5FTP, 5FWV, 5GG4, 5GKJ, 5GNI, 5GPD, 5GPJ, 5GYD, 5GYK, 5HA9, 5HBZ, 5HEV, 5HFJ, 5HIY, 5HJL, 5HMO, 5HVN, 5I6C, 5I6F, 5I6H, 5I6I, 5I8I, 5IDJ, 5IMJ, 5INE, 5IRM, 5IUM, 5J69, 5J6E, 5J6O, 5JI2, 5JJ7, 5JPZ, 5JQZ, 5JRK, 5K4L, 5K5T, 5K6W, 5K6Y, 5KH4, 5KHN,
			5KMP, 5L0G, 5L22, 5L25, 5L2E, 5L56, 5LAD, 5LF5, 5LHY, 5LLW, 5LNH, 5LOI, 5LRA, 5LZQ, 5M4R, 5MBB, 5MDM, 5MJ0, 5MN7, 5MU8, 5MZ4, 5N08, 5N6U, 5NC8, 5NYJ, 5SUQ, 5SZP, 5T1V, 5T3W, 5T9T, 5TBF, 5TGL, 5TSZ, 5TUA, 5U1G, 5U36, 5U5A, 5UCG, 5UEN, 5UFL, 5V57, 5WX5, 5WY1, 5XAR
	Homo-3-mer (trimer)	A3 (132)	1A81, 1B8H, 1BH0, 1C4T, 1EBO, 1G88, 1H6S, 1KQA, 1MPF, 1MPQ, 1MR9, 1QVR, 1QZ2, 1RER, 1RH7, 1SLQ, 1UD1, 1USZ, 1W60, 1YA8, 1ZTM, 2A81, 2AT2, 2AYN, 2B26, 2BT8, 2FKJ, 2INY, 2IU9, 2JF3, 2O0O, 2O8X, 2Q06, 2Q7M, 2R30, 2RLA, 2VDD, 2WRE, 2WST, 2XAT, 2XX1, 2YGC, 2YGQ, 2YHJ, 2YKP, 2YO1, 2ZOM, 3B29, 3BK6, 3C3I, 3D8M, 3EKA, 3EMO, 3FBA, 3FOQ, 3FYK, 3G7T, 3GDU, 3H1C, 3HON, 3IO0, 3M05, 3ME1, 3MUU, 3NWF, 3NWU, 3OER, 3OFV, 3PX1, 3QIL, 3S5C, 3SEV, 3T1P, 3T45, 3U9B, 3UBV, 3V56, 3W9J, 3X2X, 4AIM, 4AIN, 4AK3, 4AUI, 4C3K, 4C53, 4C8S, 4CM0, 4EX9, 4FI9, 4GU4, 4IDX, 4JDM, 4K0J, 4KJS, 4KY0, 4M18, 4M8J, 4MTK, 4N6B, 4NDZ, 4NYK, 4P5H, 4P7M, 4PD9, 4QKV, 4UW0, 4UXG, 4UXW, 4X6L, 4ZDD, 4ZV3, 5AEX, 5AZS, 5CH6, 5CTG, 5EGI, 5H36, 5HK2, 5I2S, 5IEA, 5JHC, 5JOT, 5K7V, 5LLU, 5LNL, 5MF1, 5SVM, 5TZ8, 5UDE, 5UN0, 5V4F, 5V68

(continued)

Table 3.1 (continued)

Low resolution (>= 3 Å) homodimer complexes			
Protein stoichiometry			PDB entries
	Homo-4-mer (tetramer)	A4 (236)	1A0D, 1A0L, 1A3W, 1A5C, 1A7B, 1AOS, 1AV1, 1AY9, 1AYR, 1AZD, 1BHW, 1BLB, 1C2O, 1C8B, 1CVI, 1DCH, 1EG1, 1FBG, 1FNV, 1FUG, 1G1R, 1GG0, 1GPH, 1GQW, 1H3Y, 1HIG, 1HN1, 1I43, 1ICT, 1JQO, 1K0U, 1M5Y, 1MSD, 1MTO, 1MV5, 1NBI, 1NIV, 1O77, 1O7Y, 1PVJ, 1QKI, 1QOJ, 1RQD, 1SFK, 1SRU, 1SZE, 1ULQ, 1V6O, 1VBP, 1WOK, 1WS3, 1WZX, 1X3G, 1XN4, 1YRH, 1YVK, 1ZCF, 2A7R, 2ABM, 2AVI, 2B4K, 2BHN, 2BJ9, 2CB6, 2DU7, 2F9Q, 2FPF, 2GRL, 2HAE, 2HCB, 2HTW, 2IX6, 2JJM, 2MSP, 2O6U, 2O94, 2OGK, 2OIG, 2PAH, 2PUW, 2Q19, 2Q5J, 2QTO, 2QVA, 2R1A, 2R5R, 2UVN, 2UVQ, 2VMK, 2VNC, 2W48, 2X0N, 2XTD, 2XTE, 2Y25, 2YQ4, 2ZII, 2ZRX, 3ALS, 3ALX, 3AMJ, 3ATD, 3AUW, 3BEH, 3C9J, 3COR, 3F6N, 3FZJ, 3GLW, 3HXK, 3HYC, 3HZQ, 3K7E, 3L7K, 3MQB, 3NAF, 3NUD, 3NY0, 3QA8, 3QLV, 3QS1, 3QSI, 3QW4, 3R2A, 3R3I, 3R6Y, 3RWT, 3SCG, 3SPH, 3T0T, 3TJ0, 3UA0, 3UG6, 3UJ3, 3UKP, 3VDR, 3VGK, 3VOU, 3WC0, 3WCB, 3ZJ4, 3ZQ4, 4AB7, 4ACQ, 4B3Z, 4BDU, 4BQL, 4BVT, 4BXT, 4CBA, 4CL1, 4CX7, 4CZ9, 4DXW, 4F0X, 4F4L, 4FIY, 4FNR, 4FQ7, 4FZH, 4GJW, 4GO1, 4GPK, 4GRS, 4GVL, 4GX2, 4H33, 4HIU, 4HNU, 4HPF, 4HYG, 4IA4, 4J8N, 4J9V, 4JJK, 4KH4, 4KSF, 4L5T, 4M9S, 4MVD, 4NSL, 4OM2, 4OVA, 4PSL, 4Q3L, 4QHH, 4QPZ, 4QSM, 4RBN, 4RH3, 4RJI, 4RZT, 4U1X, 4U4M, 4UXJ, 4W80, 4X41, 4X8A, 4XNG, 4YZS, 4ZDK, 4ZN8, 4ZSZ, 4ZXV, 5A3F, 5A9F, 5AGC, 5AO0, 5AO3, 5AXL, 5C1I, 5C31, 5C3M, 5C70, 5C7E, 5CBG, 5E38, 5EEB, 5EKE, 5ENV, 5EY2, 5FGJ, 5GGX, 5HEK, 5HHU, 5HOT, 5IP0, 5IWK, 5IWO, 5J7I, 5KCA, 5KMF, 5L6S, 5SUL, 5UQE, 5XLK, 7PCK
	Homo-5-mer (pentamer)	A5 (23)	1B44, 1C41, 1KZ9, 1LJ7, 1LTR, 2BR7, 2OAR, 2R5I, 3B8P, 3DRX, 3NWI, 3P0A, 3Q08, 3TDR, 4BTP, 4EV6, 4HFC, 4I7A, 4K7H, 4Y7K, 5D65, 5HW9, 5JTG

(continued)

Table 3.1 (continued)

Low resolution (> = 3 Å) homodimer complexes			
Protein stoichiometry			PDB entries
	Homo-6-mer (hexamer)	A6 (107)	1B26, 1DLP, 1DO0, 1E0J, 1GKI, 1HC1, 1K39, 1LNL, 1LTL, 1M8K, 1NLZ, 1OJL, 1OYR, 1RXU, 1SZP, 1T8Y, 1V1S, 1W57, 1YQU, 1Z0T, 1Z7E, 1ZXO, 2ANC, 2BRI, 2CHQ, 2DXL, 2HVV, 2OAQ, 2PG8, 2TUN, 2UY9, 2V9P, 2VHT, 2VUY, 2XAW, 3AZ8, 3B4S, 3DOI, 3EHK, 3ETE, 3FEM, 3FZ8, 3GV2, 3HWS, 3IBB, 3JVU, 3M9B, 3MH4, 3NM3, 3O6X, 3PRJ, 3Q6M, 3RCC, 3S2C, 3TIR, 3UX4, 4BWY, 4BZZ, 4D80, 4DK1, 4EO2, 4F86, 4H5M, 4HGY, 4HKR, 4HT7, 4HUH, 4IM8, 4KSR, 4LDD, 4LQL, 4LYG, 4LZZ, 4MLK, 4NMN, 4NSC, 4PYD, 4QB9, 4RBV, 4RCN, 4RHL, 4S14, 4U0D, 4U87, 4XDS, 4YPL, 4Z73, 4Z8X, 5A3U, 5AYY, 5B7B, 5C1B, 5C7P, 5CFY, 5CMD, 5DYI, 5EBZ, 5EHB, 5EXX, 5FB5, 5H5V, 5I4T, 5KC7, 5KZF, 5LKM, 5M1B, 5TSG, 5X15
	Homo-8-mer (octamer)	A8 (48)	1AHV, 1CRK, 1GYL, 1V5W, 1VCG, 1X9J, 2CHR, 2CWO, 2CXE, 2GQQ, 2HMW, 2I5K, 2PFD, 2VHI, 2W29, 2W8I, 2WIL, 2YQ0, 2ZTE, 3A8E, 3HHG, 3HVD, 3IWM, 3KFE, 3LA6, 3RBZ, 3T1E, 3T2G, 3VJJ, 3ZC1, 4AEY, 4B6J, 4B7W, 4C5Y, 4EQV, 4J90, 4M9X, 4R7P, 4RO0, 4TSY, 4WL1, 4WYV, 4YLC, 4YXY, 5ENY, 5ILA, 5K8N, 5LZL
	Homo-12-mer (dodecamer)	A12 (34)	1FOU, 1L8I, 1ORT, 1VEQ, 2GLJ, 2IJZ, 2OF5, 2W8O, 2WUY, 2Z3A, 2ZY2, 3GOK, 3LJ5, 3SOA, 3VDX, 3ZO6, 4AZV, 4CL0, 4DCL, 4DK0, 4HI2, 4LID, 4LNN, 4RB4, 4WWV, 4YNN, 4YO5, 4ZC0, 5E7S, 5ER7, 5IY0, 5JJ1, 5LFP, 5LPC
	Homo-60-mer	A60 (23)	1A6C, 1B5S, 1DNV, 1DZL, 1S58, 1VAK, 1VB4, 1VCR, 1WCD, 2C6S, 2CAS, 2ZZQ, 3DKT, 3KIE, 3MK3, 3P0S, 3ZXA, 4GBT, 4PB6, 4QC8, 4RFT, 4Y4Y, 5EGC

3.2.5 A2B2 Tetramer Protein Complex

Figure 3.8 shows the structure of a hetero tetramer (4 subunits) protein complex of myelo-peroxidase enzyme (A2B2) with PDB ID: 1MYP solved at resolution at 3 Å. It should be noted that subunits A and B (100% sequence identity) as well as C and D (100% sequence identity) are identical among themselves. There are 125 such A2B2 hetero complexes in the dataset next in number (size) to heterodimers.

3.2.6 A3B3 Hexamer Protein Complex

Figure 3.9 shows the structure of a hetero hexamer (6 subunits) of a transport protein complex (A3B3) with PDB ID: 2BSK solved at resolution at 3.3 Å. It should be noted that subunits A, C, and E representing TIM9 A are homologous among

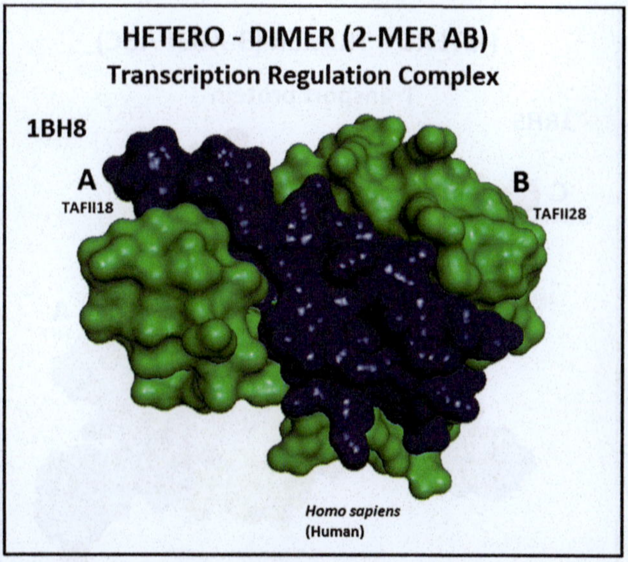

Fig. 3.4 Structure of a heterodimer (2 subunits) protein complex (AB) of a transcription regulation complex with PDB ID: 1BH8 (resolution at 3.04 Å) is drawn using Discovery Studio Visualizer™ version 16.1 [2]

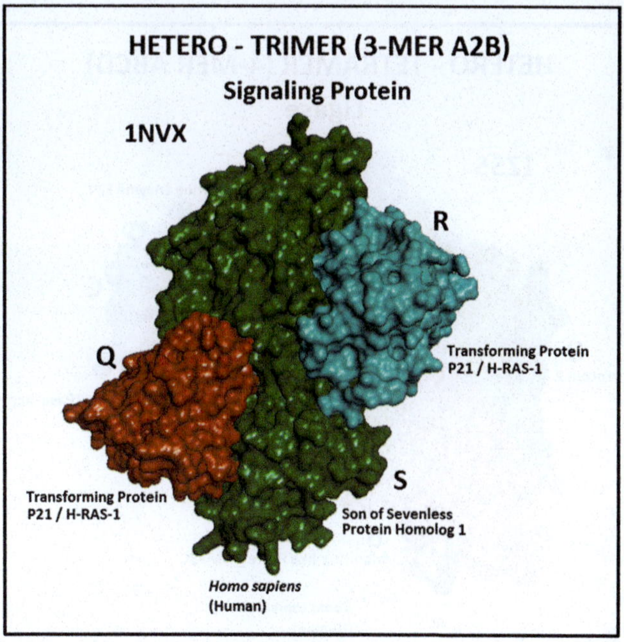

Fig. 3.5 Structure of a hetero trimer (3 subunits) protein complex (A2B) of a signaling protein with PDB ID: 1NVX (resolution at 3.2 Å) is drawn using Discovery Studio Visualizer™ version 16.1 [2]. It should be noted that 2 subunits are identical and 1 subunit is nonidentical to the other two

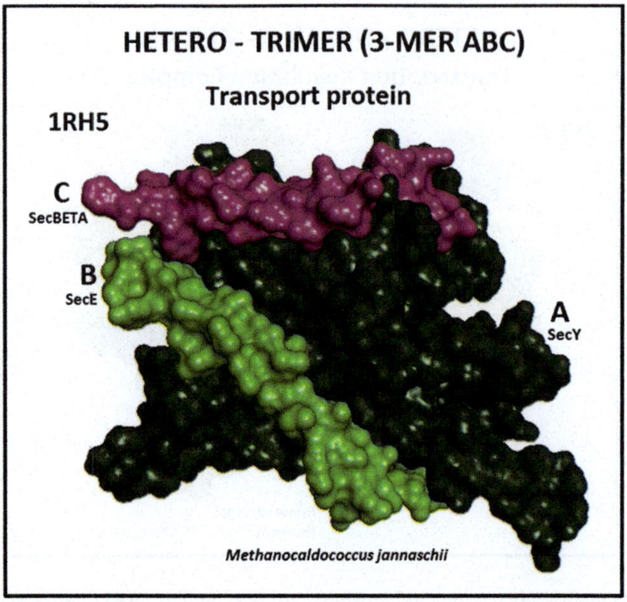

Fig. 3.6 Structure of a hetero trimer (3 subunits) transport protein complex (ABC) with PDB ID: 1RH5 (resolution at 3.2 Å) is drawn using Discovery Studio Visualizer™ version 16.1 [2]. All the three chains are non-identical in this complex

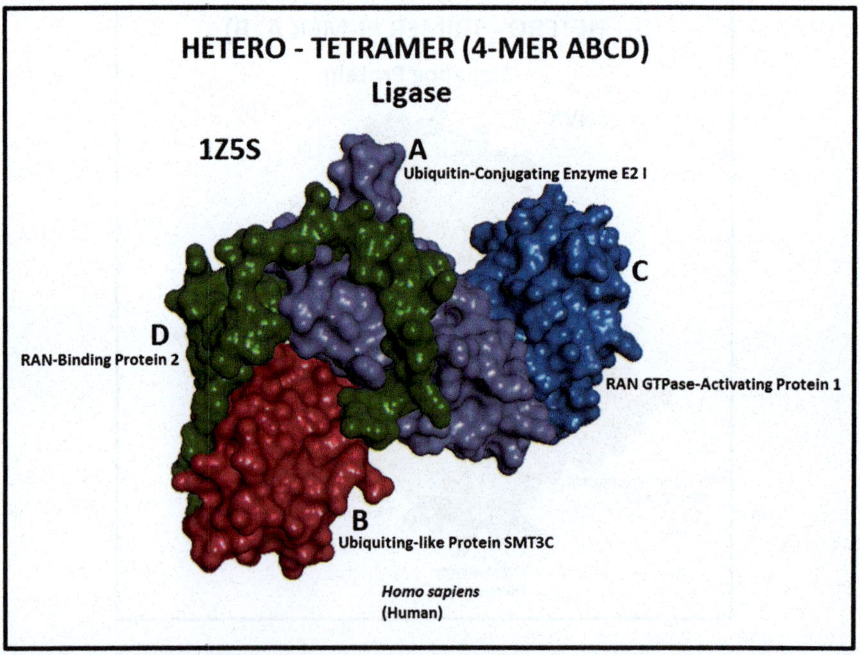

Fig. 3.7 Structure of a hetero tetramer (4 subunits) protein complex of a ligase enzyme (ABCD) with PDB ID: 1Z5S (resolution at 3 Å) is drawn using Discovery Studio Visualizer™ version 16.1 [2]. All the four subunits are non-identical in this complex

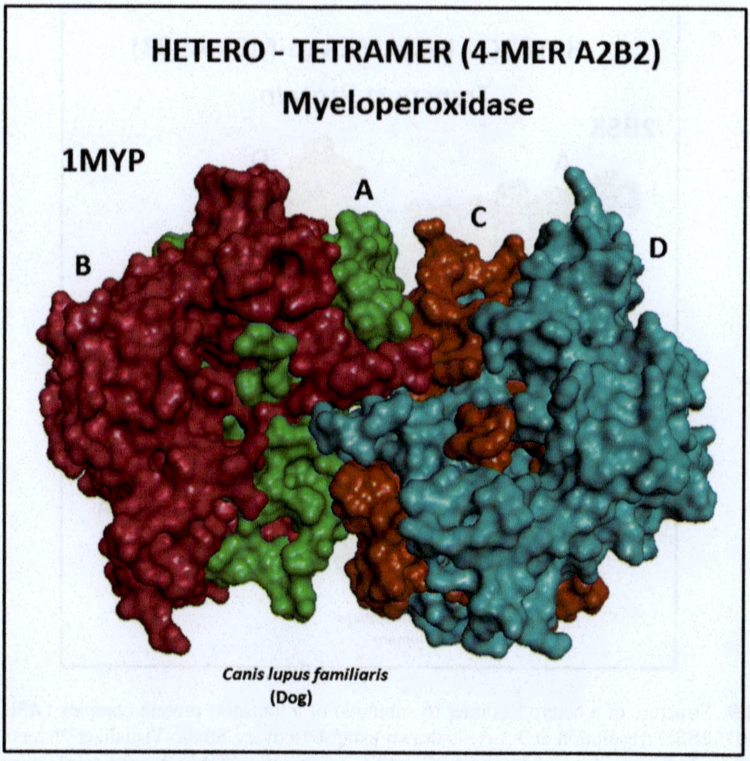

Fig. 3.8 Structure of a hetero tetramer (4 subunits) protein complex of myeloperoxidase enzyme (A2B2) with PDB ID: 1MYP (resolution at 3 Å) is drawn using Discovery Studio Visualizer™ version 16.1 [2]. It should be noted that subunits A and B (100% sequence identity) as well as C and D (100% sequence identity) are identical among themselves

themselves and subunits B, D, and F representing TIM10 are homologous among themselves. There are about 26 such A3B3 complexes in the dataset.

3.2.7 A2B2C2 Hexamer Protein Complex

Figure 3.10 shows the structure of a hetero hexamer (6 subunits) of an electron transport protein complex (A2B2C2) with PDB ID: 1O95 solved at resolution at 3.7 Å. It should be noted that A and B are homologous, C and E are homologous, and D and F are homologous. There are about 29 such A2B2C2 hetero complexes in the dataset.

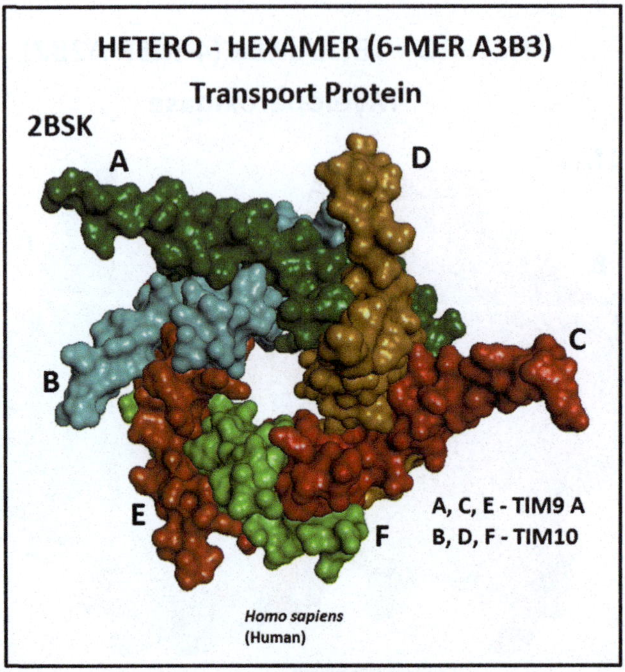

Fig. 3.9 Structure of a hetero hexamer (6 subunits) of a transport protein complex (A3B3) with PDB ID: 2BSK (resolution at 3.3 Å) is drawn using Discovery Studio Visualizer™ version 16.1 [2]. It should be noted that subunits A, C, and E representing TIM9 A are homologous among themselves and subunits B, D, and F representing TIM10 are homologous among themselves. TIM = mitochondrial import inner membrane translocase

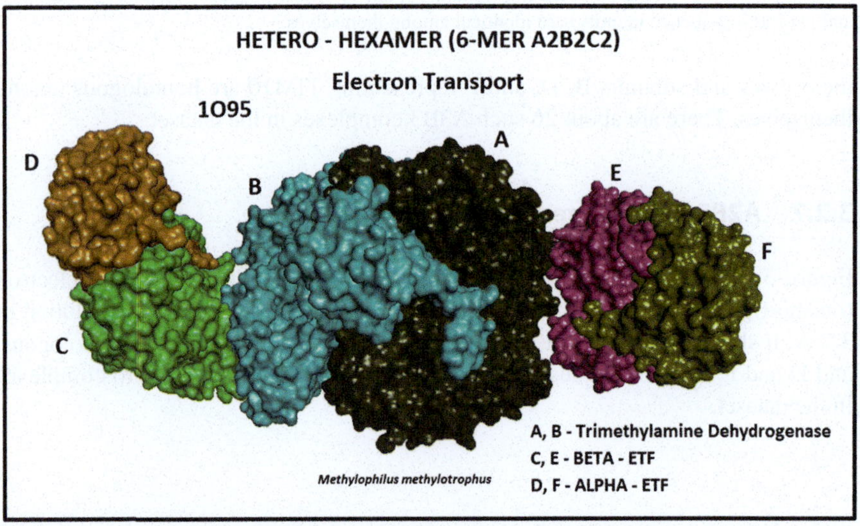

Fig. 3.10 Structure of a hetero hexamer (6 subunits) of an electron transport protein complex (A2B2C2) with PDB ID: 1O95 (resolution at 3.7 Å) is drawn using Discovery Studio Visualizer™ version 16.1 [2]. It should be noted that A and B are homologous, C and E are homologous, and D and F are homologous. ETF = electron transport flavor protein

Table 3.2 Low-resolution homodimer complexes (> = 3 Å)

Low-resolution (> = 3 Å) heteromer dataset			
Protein stoichiometry			PDB entries
Heteromer protein complexes (592)	Hetero-2-mer (dimer)	AB (271)	1BH8, 1D8E, 1DTT, 1DWC, 1FLL, 1FSS, 1G0Y, 1GH6, 1GXD, 1HRP, 1I4E, 1I8L, 1II8, 1IJF, 1IXS, 1KFX, 1KIU, 1KMH, 1LDJ, 1LPA, 1LQM, 1M10, 1MBV, 1MIU, 1MQS, 1O96, 1OGY, 1P69, 1QA9, 1RPQ, 1RY7, 1SC5, 1SGH, 1YKH, 1Z5X, 1ZT2, 1ZW3, 2A73, 2AHO, 2ATQ, 2B12, 2B3T, 2CH4, 2CKZ, 2DOI, 2DZO, 2EX3, 2F9J, 2GRX, 2GY7, 2HSM, 2J04, 2JKI, 2PF4, 2PJW, 2PM9, 2Q7N, 2RAX, 2RF4, 2UZY, 2W54, 2WG4, 2X0B, 2Y9Y, 2YSU, 2Z6K, 2ZIT, 2ZIX, 2ZUP, 3A7A, 3AV0, 3C7N, 3CQG, 3CSN, 3E0J, 3E20, 3EG9, 3ENH, 3ESW, 3EWE, 3FBN, 3FMP, 3GTY, 3GYX, 3HW2, 3I08, 3JRO, 3JUA, 3K3W, 3K5B, 3KQ4, 3L0X, 3L9K, 3LQ5, 3LWW, 3M1C, 3M5I, 3MV3, 3NVM, 3O42, 3O6B, 3OC3, 3OED, 3OGL, 3PHF, 3QB7, 3QEM, 3QMZ, 3R4D, 3REB, 3S48, 3S4W, 3S8V, 3SCJ, 3SOH, 3TBI, 3TCX, 3TU5, 3U73, 3U7U, 3ULA, 3UUX, 3V89, 3VX7, 3VX8, 3W5V, 3WEE, 3WHL, 3WKT, 3WO3, 3WOF, 3X29, 3ZEF, 3ZHE, 3ZL7, 3ZLD, 3ZXU, 3ZYJ, 4A3V, 4A5W, 4ARZ, 4ATB, 4AYM, 4BDV, 4BFI, 4BGD, 4BIK, 4BJ6, 4BOZ, 4BPX, 4BQ7, 4BRU, 4BWF, 4BXS, 4C3O, 4C5H, 4CHT, 4CV5, 4DAY, 4DX0, 4F48, 4F5C, 4FGI, 4FHM, 4FMO, 4G8X, 4H1W, 4HGK, 4HXI, 4J2L, 4JEU, 4JSN, 4KC3, 4KVO, 4L29, 4LC9, 4LGB, 4LGD, 4LI2, 4LQS, 4M4R, 4MYW, 4O9T, 4OFY, 4ORB, 4OTU, 4P00, 4P4S, 4PBW, 4PJU, 4PWV, 4Q58, 4QJ3, 4QTS, 4RHB, 4RWS, 4TU3, 4TZ4, 4U0Q, 4U1C, 4U2X, 4U39, 4U6U, 4U7P, 4UX3, 4WLP, 4WRM, 4WWX, 4WYQ, 4X63, 4XLW, 4XSS, 4XT3, 4XVP, 4XZV, 4Y61, 4Y66, 4YDO, 4YEB, 4YFD, 4YGE, 4YK8, 4YPI, 4ZGQ, 4ZIR, 4ZKT, 4ZLT, 5A7D, 5AJD, 5B2G, 5B4X, 5BQM, 5BWA, 5C8S, 5CAY, 5CHV, 5CMN, 5CYK, 5D3I, 5DO7, 5DXH, 5ECG, 5ECJ, 5EW5, 5F3J, 5F5P, 5FXY, 5G53, 5GQR, 5HVG, 5HY3, 5J3Y, 5JA1, 5JJW, 5K1B, 5L0W, 5L3G, 5L3Q, 5LHH, 5LON, 5LXQ, 5LZ3, 5MB9, 5SY5, 5T3R, 5U1S, 5U89, 5UJ8, 5UN6, 5V6E, 7API
	Hetero-3-mer (trimer)	ABC (72)	1CJV, 1FFT, 1RH5, 2ASS, 2CAZ, 2DF4, 2HJ6, 2HPQ, 2PUK, 2QMU, 2WP8, 2X1G, 3AAE, 3CJC, 3EGX, 3FEX, 3GGR, 3IKO, 3MIA, 3MK7, 3N42, 3P5B, 3PUF, 3QAZ, 3QKR, 3QQC, 3TKN, 3U88, 3VON, 3WO4, 4C0D, 4CFE, 4DEP, 4DGW, 4EJX, 4F52, 4FXK, 4G7R, 4GK5, 4H62, 4HYT, 4MMZ, 4MN8, 4MNG, 4NKQ, 4OWR, 4R3Z, 4R71, 4UFS, 4UJ3, 4UWQ, 4Y49, 4YCZ, 5AFF, 5BUZ, 5CH4, 5CVN, 5D98, 5DOQ, 5F0M, 5F5V, 5FO9, 5FWW, 5HE1, 5HXB, 5J12, 5L3T, 5LOP, 5LS6, 5N2V, 5N5X, 5UJ7

(continued)

Table 3.2 (continued)

Low-resolution (>= 3 Å) heteromer dataset			
Protein stoichiometry			PDB entries
	Hetero-3-mer (trimer)	A2B (41)	1NVX, 1U2D, 1UKL, 1Z56, 2GV5, 2IBF, 2XCN, 3AAD, 3EIQ, 3ENO, 3EUK, 3F5C, 3F9K, 3IKM, 3LTG, 3M3N, 3NPZ, 3SJD, 3T53, 3U3F, 3W5A, 4DKD, 4EBA, 4GU0, 4HSA, 4LL8, 4LXS, 4N3Z, 4RG6, 4UHY, 4URT, 4XR7, 5B0O, 5B16, 5B1J, 5EJC, 5IUL, 5JHQ, 5KTB, 5LJY, 5NBD
	Hetero-4-mer (tetramer)	A2B2 (125)	1AIP, 1CQI, 1EUI, 1FQ9, 1H4J, 1HLT, 1I7X, 1I85, 1IGY, 1JCH, 1JRP, 1LEM, 1MAH, 1MIO, 1MYP, 1OQL, 1PGR, 1RHR, 1VIW, 1VYH, 1W1I, 1WDL, 1WDW, 1ZQ1, 2BQ1, 2CG9, 2D0P, 2E9W, 2F9Y, 2FQQ, 2GID, 2H4M, 2IFG, 2NZV, 2O8V, 2P5T, 2POP, 2Q2E, 2XZ1, 2YJN, 2ZBK, 2ZVN, 3AL8, 3B13, 3CL3, 3D0H, 3D12, 3D31, 3EB6, 3EUJ, 3EXF, 3F3P, 3FXI, 3IJ2, 3INB, 3KIN, 3KYH, 3L4G, 3LVL, 3M51, 3MHV, 3NFF, 3NMZ, 3NUH, 3O1H, 3O4X, 3Q35, 3Q4F, 3QG5, 3QS7, 3S9N, 3SJA, 3TUZ, 3UWX, 3V65, 3WKS, 3ZGX, 4APF, 4B7Y, 4BKF, 4BKX, 4C31, 4C9U, 4CZY, 4DGL, 4EAH, 4G1U, 4HUW, 4JAS, 4JO8, 4LPZ, 4LRX, 4P69, 4P73, 4PE5, 4PLN, 4RA0, 4RND, 4S0T, 4UMO, 4UX6, 4WPB, 4WYK, 4YCG, 4YMV, 4YTX, 5AKB, 5AUP, 5B6B, 5BQC, 5CIB, 5CL1, 5D7G, 5DA9, 5DBO, 5E9U, 5FM5, 5IC9, 5IFW, 5K22, 5LN1, 5NXB, 5T4Y, 5UHL, 5V6T
	Hetero-4-mer (tetramer)	ABCD (28)	1M57, 1Z5S, 2BCJ, 2EHO, 2ERJ, 2PVO, 3D38, 3HD7, 3HS0, 3UAI, 3VH6, 4CFH, 4JGH, 4OGR, 4P6Z, 4YTN, 5C3I, 5DJQ, 5EKQ, 5EXR, 5FU7, 5HCE, 5ICA, 5IFE, 5J9Q, 5JK7, 5LSJ, 5T58
	Hetero-6-mer (hexamer)	A2B2C2 (29)	1E7P, 1FZI, 1JJO, 1LGB, 1O95, 1P1Z, 1P9M, 2P2C, 2PJY, 2VPW, 3FM7, 3RNA, 4DJF, 4IIT, 4O9U, 4PWX, 4Q66, 4TMA, 4TRQ, 4WHV, 4WPX, 4XRP, 5BS7, 5BWE, 5C2W, 5CHE, 5JHF, 5KLI, 5KQ1
	Hetero-6-mer (hexamer)	A3B3 (26)	1FAV, 1FLC, 1IJD, 1PK0, 1RF3, 1SKY, 2BSK, 2C38, 2RDD, 2Z6B, 3EN1, 3HMJ, 3L89, 3LVG, 3P83, 3PXI, 3VR3, 4FZ1, 4JPB, 4JUJ, 4MMU, 4ODB, 4ZMJ, 5CIR, 5L36, 5U68

3.3 Protein Homo Complexes

There are about 1500 low-resolution non-redundant homo complexes (Table 3.2) at the protein data bank (PDB) as shown in Fig. 3.2. The distribution of these complexes based on the number of subunits is shown in Fig. 3.3. Homodimers (A2) are the most represented protein complex structures (860) in the dataset. The second most represented structures (236) are A4 homo tetramers.

3.3.1 A2 Homodimer Protein Complex

Figure 3.11 shows the structure of a homodimer (2 subunits) DNA-binding protein complex with PDB ID: 1ADU solved at resolution at 3 Å. It should be noted that both the subunits are identical and homologous in this complex. There are about 860 A2 dimer complexes in the dataset and it is the most represented.

3.3.2 A3 Homo Trimer Protein Complex

Figure 3.12 shows the structure of a homo trimer (3 subunits) of a transcription protein complex with PDB ID: 1G88 solved at resolution at 3 Å. It should be noted that all the 3 subunits are identical and homologous in this complex. There are about 132 homo A3 trimer complexes in the dataset. This is interesting.

3.3.3 A4 Homo Tetramer Protein Complex

Figure 3.13 shows the structure of a homo tetramer (4 subunits) human beta tryptase protein complex (A4) with PDB ID: 1A0L solved at resolution at 3 Å. It should be noted that all the 4 subunits are homologous at 100% sequence identity. This is the

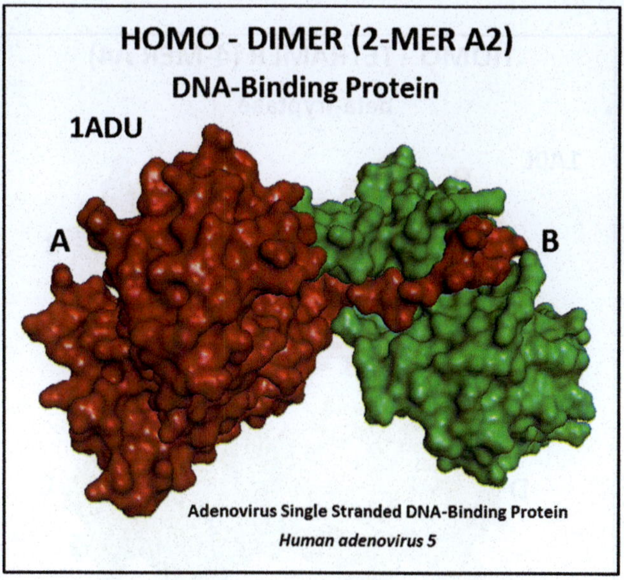

Fig. 3.11 Structure of a homodimer (2 subunits) DNA-binding protein complex with PDB ID: 1ADU (resolution at 3 Å) is drawn using Discovery Studio Visualizer™ version 16.1 [2]. It should be noted that both the subunits are identical and homologous in this complex

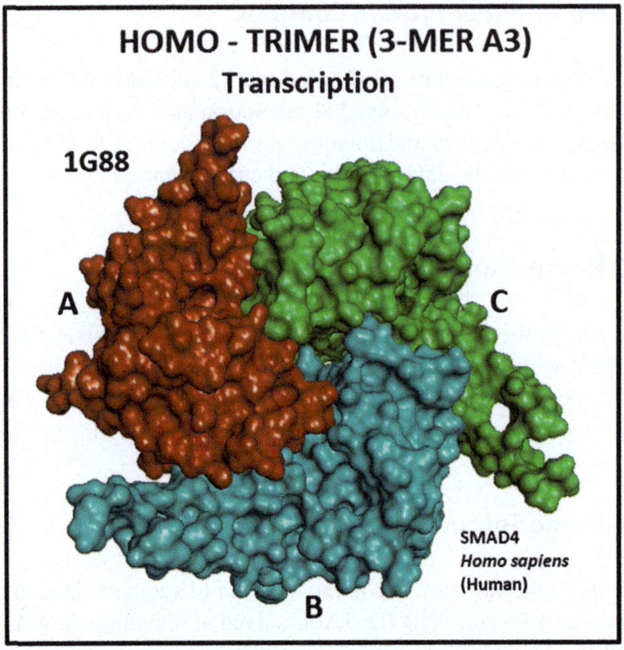

Fig. 3.12 Structure of a homo trimer (3 subunits) of a transcription protein complex with PDB ID: 1G88 (resolution at 3 Å) is drawn using Discovery Studio Visualizer™ version 16.1 [2]. It should be noted that all the 3 subunits are identical and homologous in this complex

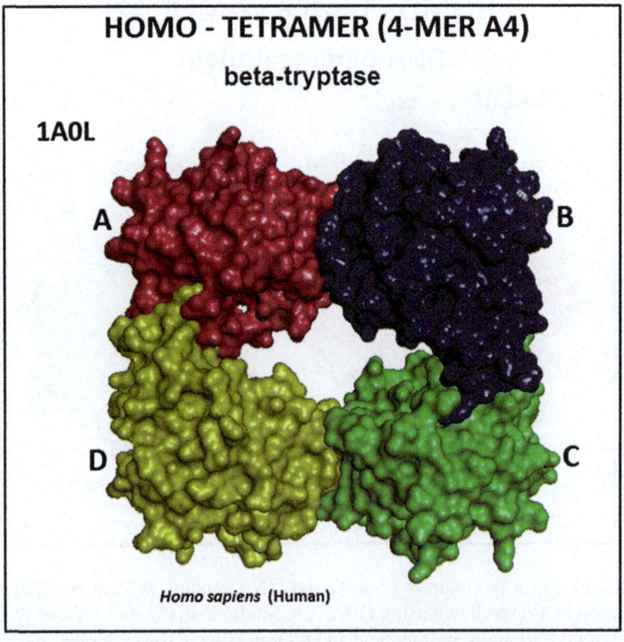

Fig. 3.13 Structure of a homo tetramer (4 subunits) human beta tryptase protein complex (A4) with PDB ID: 1A0L (resolution at 3 Å) is drawn using Discovery Studio Visualizer™ version 16.1 [2]. It should be noted that all the 4 subunits are homologous at 100% sequence identity

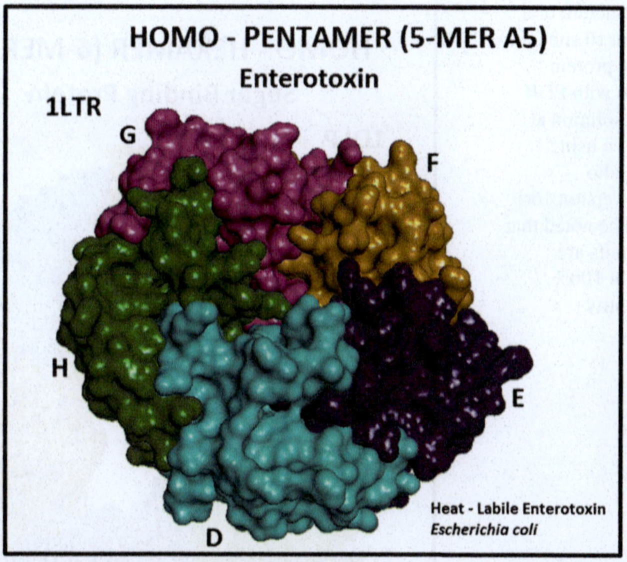

Fig. 3.14 Structure of a homo pentamer (5 subunits) enterotoxin protein complex (A5) with PDB ID: 1LTR (resolution at 3 Å) is drawn using Discovery Studio Visualizer™ version 16.1 [2]. It should be noted that all the 5 subunits are homologous at 100% sequence identity

second most represented structures in the dataset. There are about 236 such A4 complexes in the dataset.

3.3.4 A5 Homo Pentamer Protein Complex

Figure 3.14 shows the structure of a homo pentamer (5 subunits) enterotoxin protein complex (A5) with PDB ID: 1LTR solved at resolution at 3 Å [2]. It should be noted that all the 5 subunits are homologous at 100% sequence identity. There are about 23 such A5 complexes in the dataset.

3.3.5 A6 Homo Hexamer Protein Complex

Figure 3.15 shows the structure of a homo hexamer (6 subunits) sugar-binding protein complex (A6) with PDB ID: 1DLP solved at resolution at 3.3 Å. It should be noted that all the 6 subunits are homologous at 100% sequence identity. There are about 107 such A6 complexes in the dataset.

Fig. 3.15 Structure of a homo hexamer (6 subunits) sugar-binding protein complex (A6) with PDB ID: 1DLP (resolution at 3.3 Å) is drawn using Discovery Studio Visualizer™ version 16.1 [2]. It should be noted that all the 6 subunits are homologous at 100% sequence identity

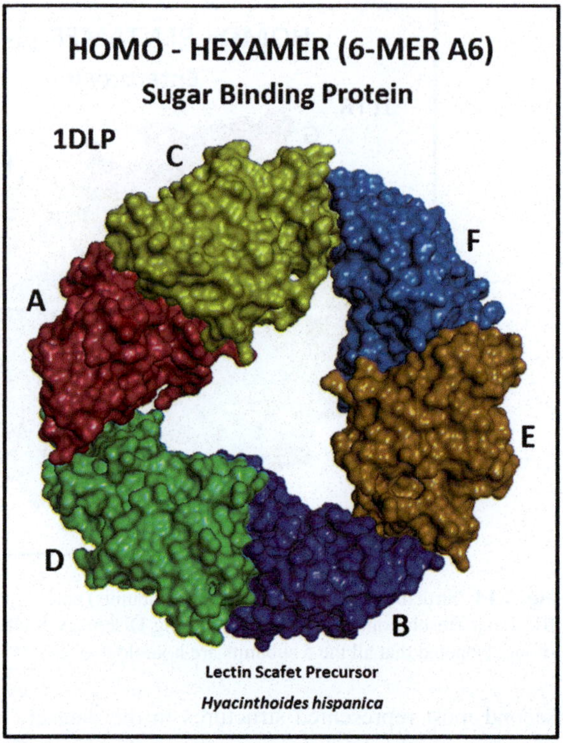

Fig. 3.16 Structure of a homo octamer transferase (butyrate kinase) protein complex (A8) with PDB ID: 1X9J (resolution at 3 Å) is drawn using Discovery Studio Visualizer™ version 16.1 [2]. It should be noted that all the 8 subunits are homologous at 100% sequence identity

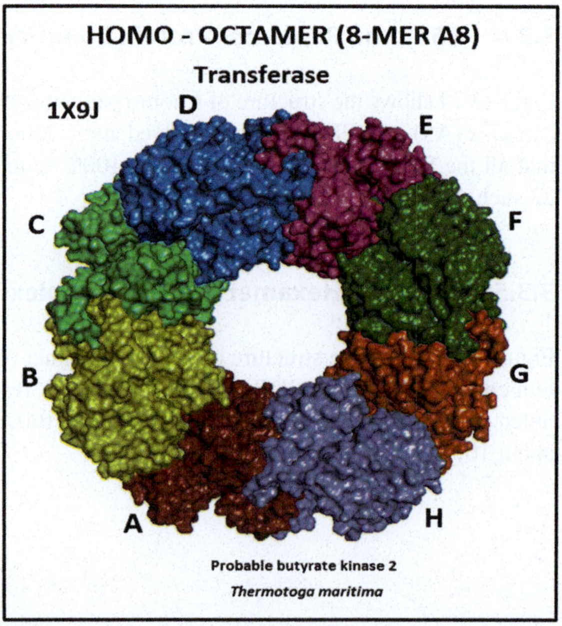

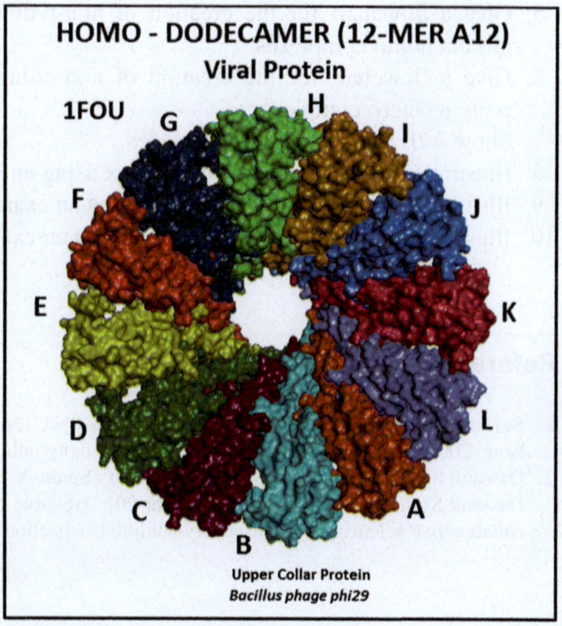

Fig. 3.17 Structure of a homo dodecamer (12 subunits) viral protein (upper collar phage phi29) complex with PDB ID: 1FOU (resolution at 3.2 Å) is drawn using Discovery Studio Visualizer™ version 16.1 [2]. It should be noted that all the 12 subunits are homologous with 100% sequence identity

3.3.6 A8 Homo Octamer Protein Complex

Figure 3.16 shows the structure of a homo octamer transferase (butyrate kinase) protein complex (A8) with PDB ID: 1X9J solved at resolution at 3 Å. It should be noted that all the 8 subunits are homologous at 100% sequence identity. There are about 48 such A8 complexes in the dataset.

3.3.7 A12 Homo Dodecamer Protein Complex

Figure 3.17 shows the structure of a homo dodecamer (12 subunits) viral protein (upper collar phage phi29) complex with PDB ID: 1FOU solved at resolution at 3.2 Å. It should be noted that all the 12 subunits are homologous with 100% sequence identity. There are about 34 such A12 complexes in the dataset.

Exercises

1. What percentage of entries in PDB are low-resolution protein complexes?
2. Illustrate a homo tetramer complex with an example.
3. Differentiate between A2B and ABC protein complexes with an example.
4. Give a flowchart for the creation of non-redundant low-resolution protein-protein complexes.

5. Give a flowchart for the creation of non-redundant low-resolution protein-protein homo complexes.
6. Give a flowchart for the creation of non-redundant low-resolution protein-protein hetero complexes.
7. Show A2B2C2 and A3B3 complexes.
8. Illustrate a dodecamer protein complex using an example.
9. Illustrate a homo octamer complex using an example.
10. Illustrate a homo pentamer complex using an example.

References

1. Berman HM, Westbrook J, Feng Z, Gilliland G, Bhat TN, Weissig H et al. The protein data bank. 2000. [cited: 2017 Mar 20]. http://www.rcsb.org/pdb/home/home.do
2. Dassault Systemes BIOVIA, BIOVIA Discovery Studio Visualizer, v16.1.0.15350, San Diego: Dassault Systemes; 2015 [cited: 2017 Mar 20]. Available from: http://accelrys.com/products/collaborative-science/biovia-discovery-studio/visualization-download.php

Heterodimer Protein Complexes 4

4.1 Heterodimer Protein Complex

A heterodimer protein complex is made up of two dissimilar protein subunits as shown in Fig. 4.1. They vary in both size and sequence similarity. The different regions (surface, core, and interface) of the complex are shown. Surface regions are solvent accessible and core are buried (solvent inaccessible). Interface is that part of the surface in monomer proteins that are buried (solvent inaccessible) upon complex formation.

4.2 Interface Size

Interface size is the number of residues in the interface that are involved in protein-protein binding. The frequency of distribution of interface size in heterodimer complexes is shown in Fig. 4.2. Majority of complexes are having interfaces with sizes of 40 to 80 residues among 192 protein complexes.

4.3 Interface Area and Interface Size

Interface area is the area buried upon complex formation. It is calculated as the change in accessible surface area (delta ASA) upon complex formation. The relationship between interface size (number of interface residues) and interface area is shown in Fig. 4.3. They are related with high correlation coefficient.

P. Kangueane, C. Nilofer, *Protein-Protein and Domain-Domain Interactions*,
https://doi.org/10.1007/978-981-10-7347-2_4

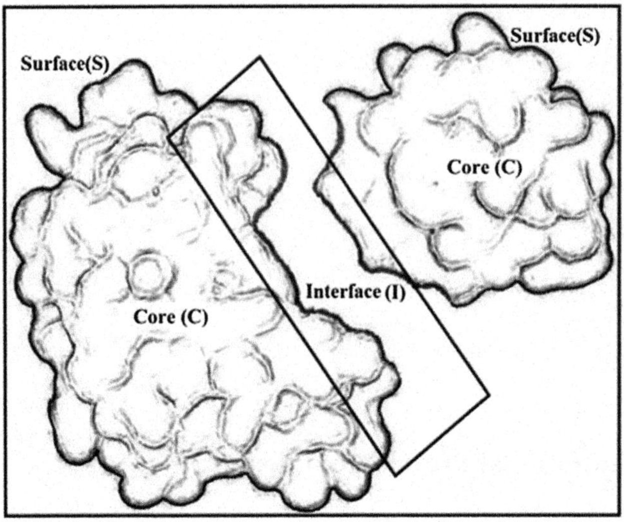

Fig. 4.1 An illustration of a heterodimeric protein complex is illustrated. The interacting proteins are dissimilar in size and sequence. The different regions (surface, core, and interface) of the complex are shown. Surface regions are solvent accessible and core are buried (solvent inaccessible). Interface is that part of the surface in monomer proteins that are buried (solvent inaccessible) upon complex formation (This image is adapted from Sowmya et al. [1] under the open access creative commons attribution license)

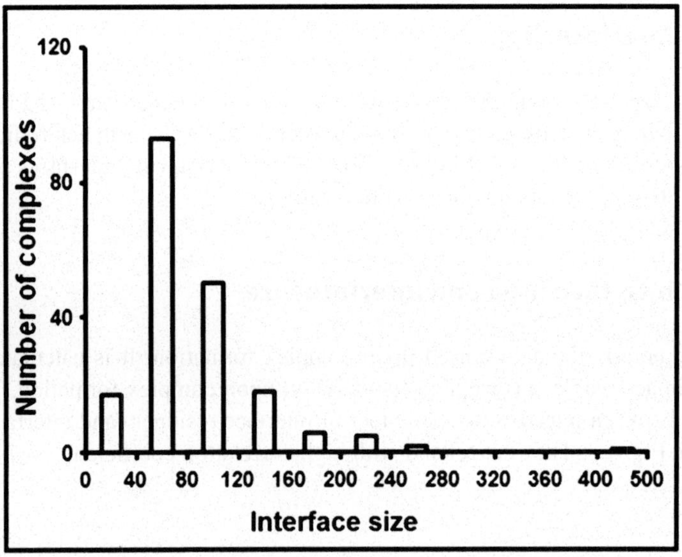

Fig. 4.2 The frequency of distribution of interface sizes in heterodimer complexes is shown. Majority of complexes are having interfaces with sizes of 40 to 80 residues among 192 protein complexes (This image is adapted from Sowmya et al. [1] under the open access creative commons attribution license)

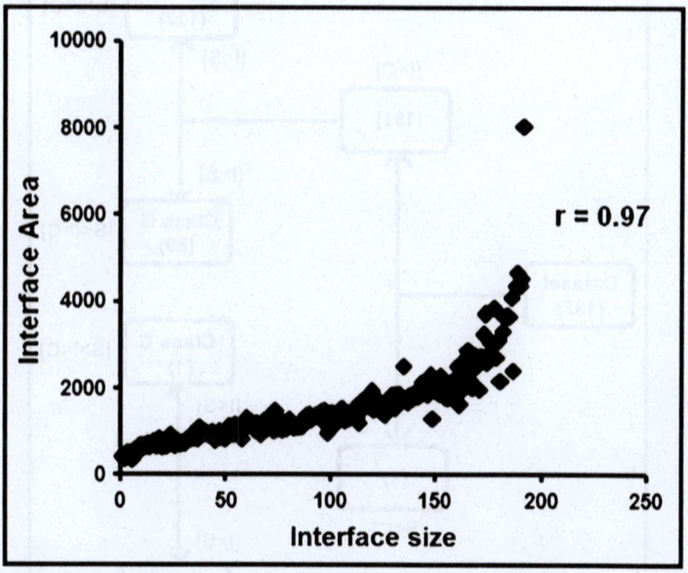

Fig. 4.3 The relationship between interface size (number of interface residues) and interface area is shown among 192 heterodimer complexes. The relation between is high with a correlation coefficient of 0.92 (This image is adapted from Sowmya et al. [1] under the open access creative commons attribution license)

4.4 Inter-residue Analysis of Complexes

Residue level analysis of protein-protein interfaces shows that interfaces are rich in either nonpolar (class A) or polar (class B) residues compared to surface as shown in Fig. 4.4.

4.5 Nonpolar and Polar Interfaces

The distribution of protein-protein complexes with either nonpolar (class A) or polar (class B) residue dominance compared to surface is shown in Fig. 4.5. Interfaces are usually rich in nonpolar residues (class A). However, a considerable number of interfaces are rich in polar residues (class B).

4.6 Protein Complexes and Molecular Function

The grouping of protein complexes based on known molecular function is shown in Fig. 4.6. They are either obligatory (essential) or non-obligatory (nonessential) in nature. Obligatory complexes are enzymes, regulators, and biological assemblies.

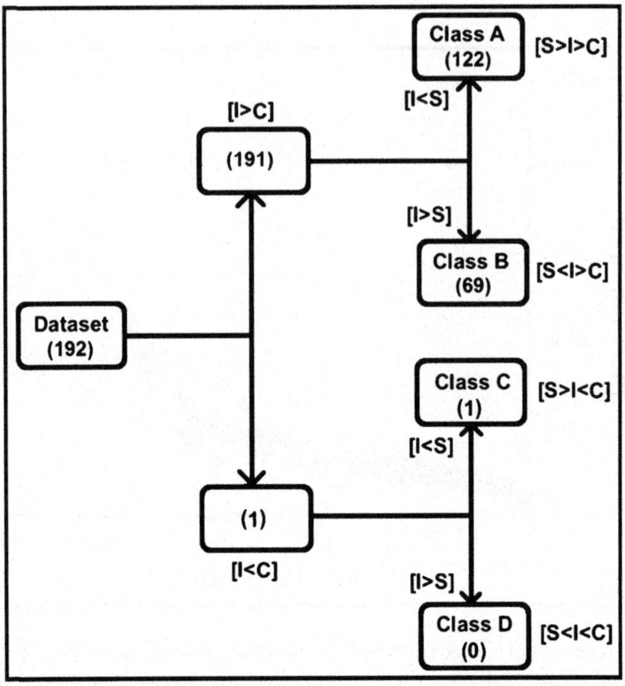

Fig. 4.4 The classification of complexes based on interfaces with abundant polar residue is shown in relation to interface (I), surface (S), and core (C). It should be noted that the core is always nonpolar in nature. The relative difference in nonpolar features between surface and interface is observed, and it is found to be interesting. Residue level analysis shows that interfaces are either rich in nonpolar (class A) or polar (class B) (This image is adapted from Sowmya et al. [1] under the open access creative commons attribution license)

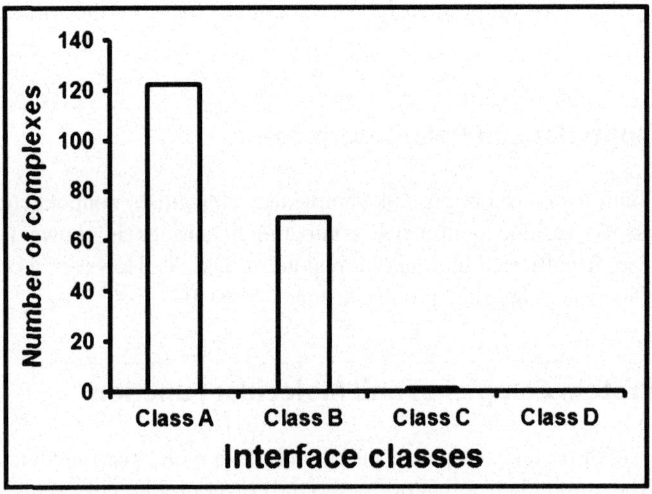

Fig. 4.5 The distribution of protein-protein complexes with either nonpolar (class A) or polar (class B) residue dominance compared to surface is shown (This image is adapted from Sowmya et al. [1] under the open access creative commons attribution license)

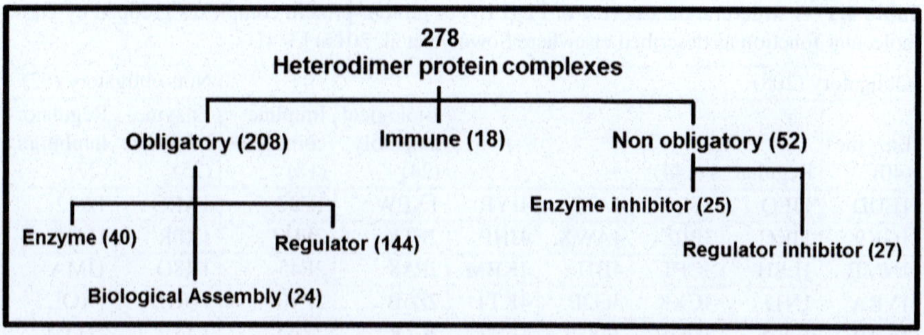

Fig. 4.6 The grouping of protein complexes based on known molecular function. They are either obligatory (essential) or non-obligatory (nonessential) in nature. Obligatory complexes are enzymes, regulators, and biological assemblies. Non-obligatory complexes are either enzyme-inhibitors or regulator-inhibitors (This image is adapted from Nilofer et al. [2] under the open access creative commons attribution license)

Non-obligatory complexes are either enzyme-inhibitors or regulator-inhibitors. This is based on the dataset of complexes given in Table 4.1. Examples of obligatory, immune, and non-obligatory complexes with known structures are shown in Fig. 4.7.

4.7 Nonpolar and Polar Interfaces with Known Molecular Function

Grouping of complexes based on interface residue features (nonpolar abundance (class A) or otherwise (class B)) with known molecular function is shown in Fig. 4.8. There is no bias in molecular function between class A and class B protein-protein complexes.

4.8 Interface Size, Energy, and Function

Protein-protein binding is proportional to interface size (Fig. 4.2) and hence with interface area (Fig. 4.3). Large interface size has large interface area and high energy (sum of van der Waals, H bonds, and electrostatics) irrespective of molecular function. This is true for obligatory, non-obligatory, and immune complexes as shown in Fig. 4.9. This is also true within obligatory complexes (Fig. 4.10) such as enzymes, regulators, and biological assembly as well as non-obligatory complexes (Fig. 4.11) such as enzyme-inhibitors and regulator-inhibitors. A similar observation is found with nonpolar interfaces (class A) as well as polar interfaces (class B) as shown in Fig. 4.12. The coefficient of determination is high between interface size and total energy as given in Tables 4.4 and 4.5.

Table 4.1 A structural dataset (list of PDB ID) of protein-protein complexes grouped by known molecular function as described elsewhere Sowmya et al. 2015) [3–4]

Obligatory (208)							Non-obligatory (52)	
Enzymes (40)	Regulators (144)				Biological assembly (24)	Immune complexes (18)	Enzyme-inhibitors (25)	Regulator-inhibitors (27)
1EUD	1JEQ	3B0C	4AT7	4IYP	1XEW	1Z92	1ARO	1A22
1GK9	1JKG	3B0Z	4AWX	4JHP	2FTX	2FHZ	1R0R	1M1E
1NME	1LSH	3CPT	4BJJ	4KBM	2PA8	2P45	1R8O	1JMA
1VRA	1N1J	3CX8	3O2P	4KT1	2Z5B	2PTT	1UGH	1XOU
1KA9	1OF5	3DGP	3OG6	1FS0	3CQC	2Z64	1AY7	2FOM
1SVD	1O00	3DPL	3OJM	1H32	3FPN	3DI3	1WPX	2HDI
2APO	1ORY	3EGV	3SBT	1NRJ	3K1R	3DLQ	1XG2	2F4M
2BLF	1US7	3CLS	3THO	1Z0J	3KCP	3FPU	2B42	2OMZ
2CG5	1WQJ	3D3B	3TU3	1Z5Y	3KF6	3MJ7	2OZN	2QC1
2GA9	1YKH	3EI3	3VF0	2GSK	3LQC	3PV6	3CKI	3N40
2XFG	2D74	3F6Q	4C9B	2PQN	3VYR	3RNQ	3DBO	2RAW
2ZIV	1DJ7	3FMO	4CBU	2V6X	3VU9	3WA5	3IEY	2VSM
2O2V	1H2V	3FQD	4CGY	3AQF	3W8I	4HFF	3QN1	3NVN
2QKL	1USU	3GB8	4CT0	3BS5	4AP2	4J38	2O3B	3ONA
3A8G	1WMH	3F62	4DEY	4CXF	4F48	4K12	2VLQ	3AA7
3AON	1Z3E	3H7H	4E4W	4G1M	4H4K	4KT3	3R24	3BTP
3AYH	1ZBX	3HZH	4EGC	1ZHH	4HPL	1V74	3N4I	3OQ3
3DRA	2CKL	3IF8	4ETP	2FCW	4IU2	1T0P	3SHG	3QQ8
3EP6	2DYO	3K8P	4EUK	2V8S	1GL4		4F6U	3NYB
3FGR	2FH5	3KXC	4EYY	3MP7	1NPE		3FJU	3OJA
3GA9	2G2S	3LF4	4G6T	3NY7	3A4U		4DRI	4BI8
3JTQ	2H9A	3MCB	3VRD	3OUN	3C5X		4F9C	4GAF
3DSS	2HRK	3MKR	3W9C	3V8X	3LBX		4HT3	4B8A
3MWD	2IW5	3MXN	3ZNZ	4G7X	3M1C		4I1S	4G94
3O3O	2VDB	3NW0	3ZYI				4LV5	4GVB
4EMJ	2P1M	3GC3	4BL7					1T6B
3L91	2QSF	3HHM	4C2A					1E44
3M7F	2QWO	3KLD	4DVG					
3ML1	2V3B	3KYJ	4GDX					
3R07	2VN6	3MCA	4JE3					
3REQ	2WD5	3N1M	4JEH					
3ZVQ	2ZFD	3NV0	4KHA					
4BMP	2ZSI	3OSS	4KMO					
4DBG	3ANW	3PGE	4L2I					
4GED	3AWU	3Q87	4M69					
4HNX	3AXJ	3T5X	4M6W					
4HST	3BEG	3TBI	4NQW					
4MRT	3A2F	3VZ9	4O8Y					
4NFU	3ABE	3ZET	4FZV					
3RGW	3AU4	4A5U	4GQ2					

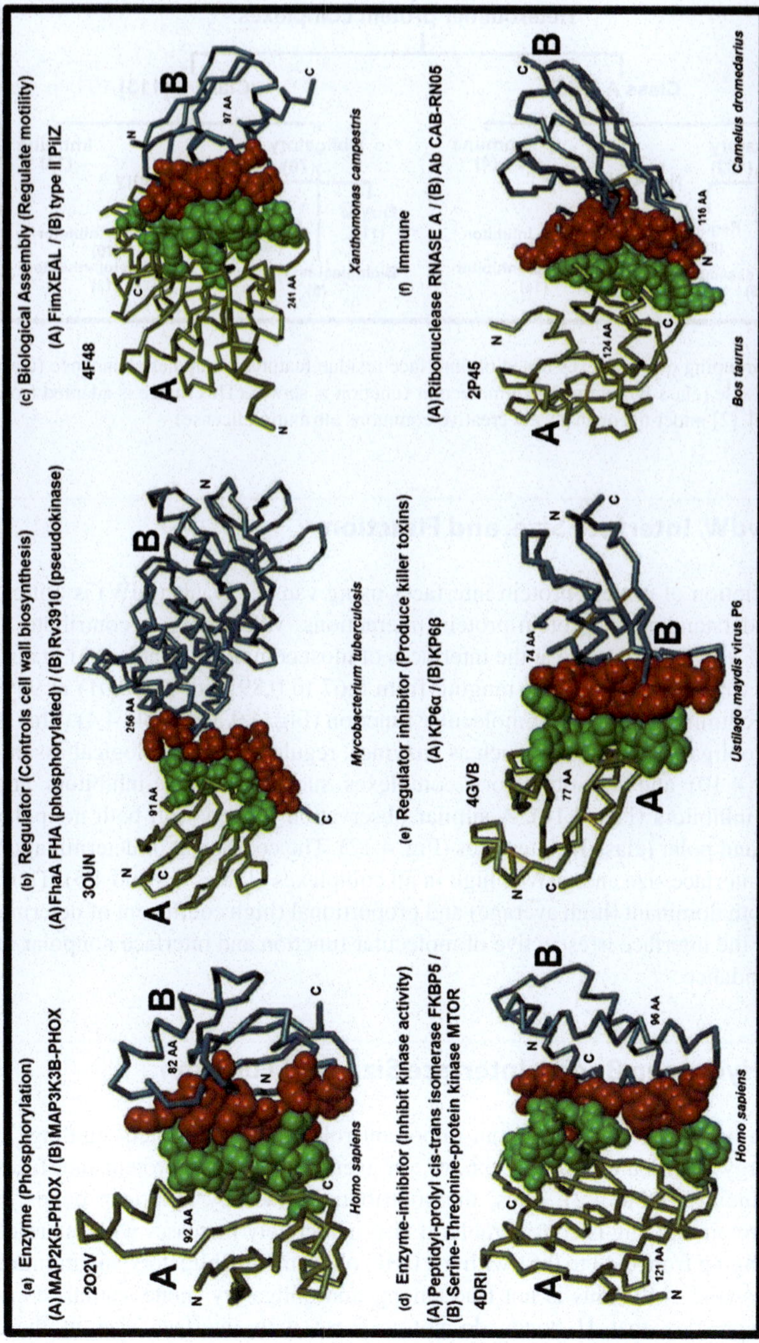

Fig. 4.7 Examples of obligatory, immune, and non-obligatory complexes with known structures are shown. **(a)** Enzyme complex, **(b)** regulator complex, **(c)** biological assembly, **(d)** enzyme-inhibitor, and **(e)** regulator-inhibitor and immune complex are illustrated using examples (This image is adapted from Nilofer et al. [2] under the open access creative commons attribution license)

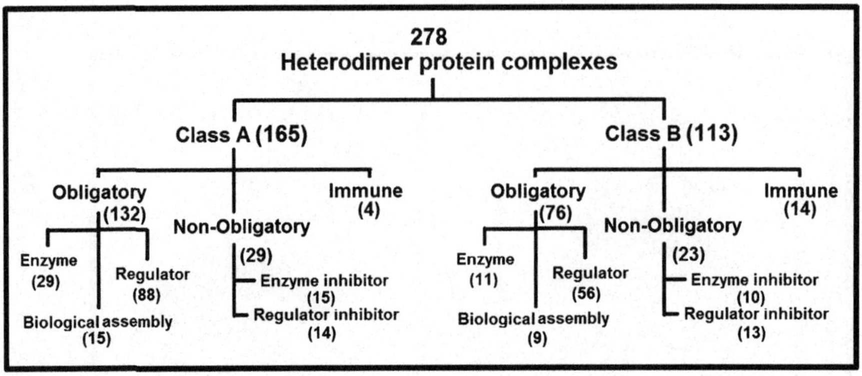

Fig. 4.8 Grouping of complexes based on interface residue features (nonpolar abundance (class A) or otherwise (class B)) with known molecular function is shown (This image is adapted from Nilofer et al. [2] under the open access creative commons attribution license)

4.9 vdW, Interface Size, and Function

The description of protein-protein interfaces using van der Waals (vdW) is critical in the understanding of protein-protein interactions. vdW energies contribute to about 75% ±11% on average at the interfaces of most complexes (Table 4.2). It also increases with interface size (r^2 ranging from 0.67 to 0.89 with $p < 0.01$) at 95% confidence limit irrespective of molecular function (Fig. 4.9 and Table 4.4). This is true with obligatory complexes such as enzymes, regulators, and biological assembly (Fig. 4.10) and non-obligatory complexes such as enzyme-inhibitors and regulator-inhibitors (Fig. 4.11). A similar observation is seen with both nonpolar (class A) and polar (class B) interfaces (Fig. 4.12). The coefficient of determination between interface size and vdW is high in all complexes (Tables 4.4 and 4.5). Thus, vdW is both dominant (high average) and proportional (high coefficient of determination) at the interface irrespective of molecular function and interface nonpolar or polar abundance.

4.10 Hydrogen Bonds, Interface Size, and Function

Hydrogen bonds (H bonds) play an important role in protein-protein binding. H bond energy contributes to 15% ± 6.5% on average in protein-protein interfaces toward binding (Table 4.2). Thus, the contribution made by hydrogen bonds in protein-protein binding is considerable. It also moderately increases with interface size (r^2 ranging from 0.43 to 0.61 with $p < 0.01$) only among obligatory and immune complexes (Fig. 4.9). This is not true among non-obligatory (nonessential) complexes suggesting that H bonds do not increase with interface size in these

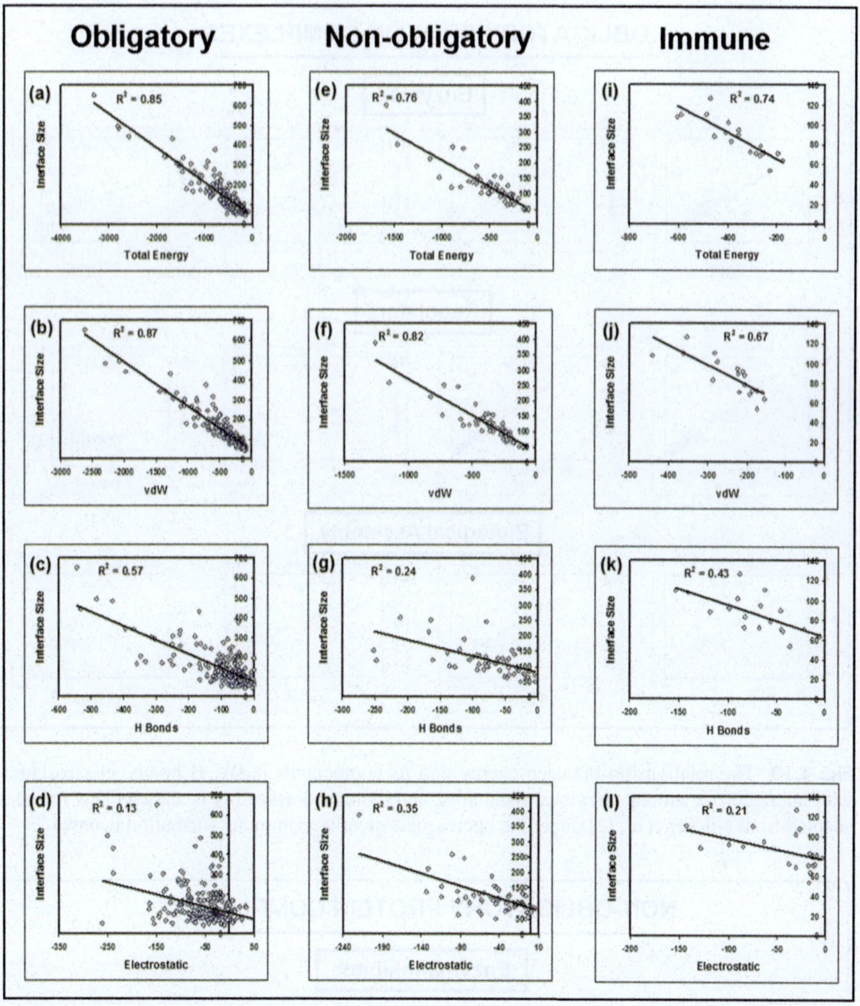

Fig. 4.9 The relationship between interface size and energy, vdW, H bonds, and electrostatics is shown for obligatory, immune, and non-obligatory complexes (This image is adapted from Nilofer et al. [2] under the open access creative commons attribution license)

complexes. The role by H bonds (14.83% ± 6.5% (Table 4.3) and $r^2 = 0.61$ with $p < 0.01$) among obligatory within interfaces (class A) having more nonpolar residues than surface is also observed. This is interesting and thought provoking. In an example complex, an obligatory regulatory protein transport (PDB: 3B0Z) interface is made of vdW (~72%) and H bonds (~28%) without electrostatics (Table 4.6). The role played by hydrogen bonds for specificity in this case is critical.

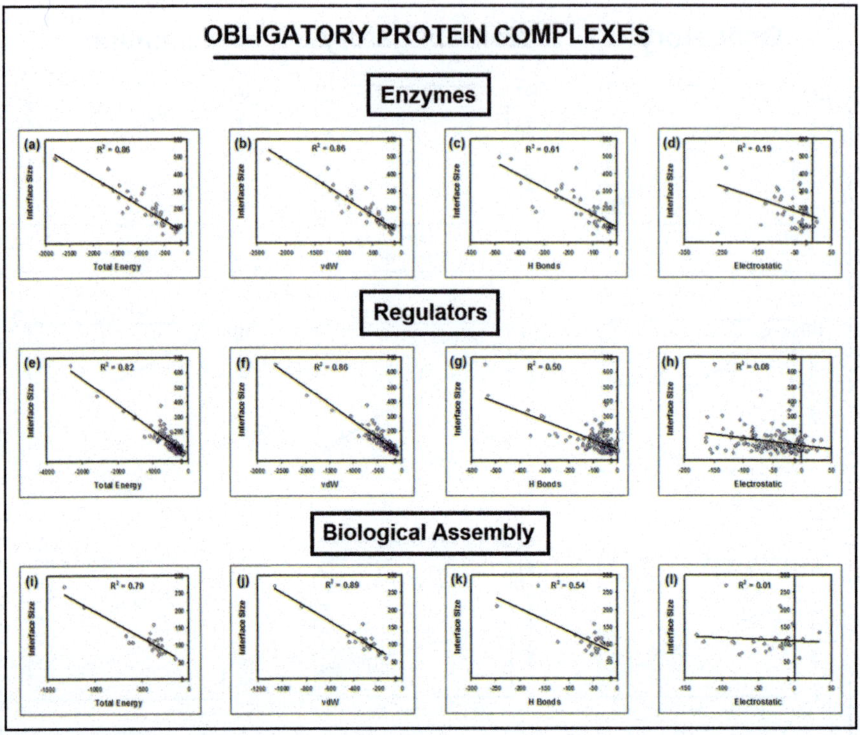

Fig. 4.10 The relationship between energy and its components (vdW, H bonds, electrostatics) with interface size among enzymes, regulators, and biological assembly is shown (This image is adapted from Nilofer et al. [2] under the open access creative commons attribution license)

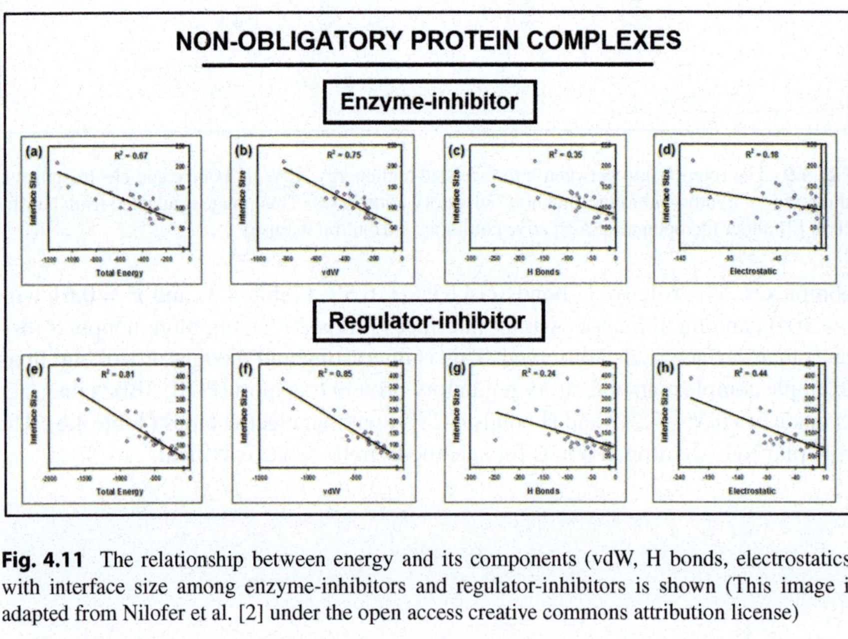

Fig. 4.11 The relationship between energy and its components (vdW, H bonds, electrostatics) with interface size among enzyme-inhibitors and regulator-inhibitors is shown (This image is adapted from Nilofer et al. [2] under the open access creative commons attribution license)

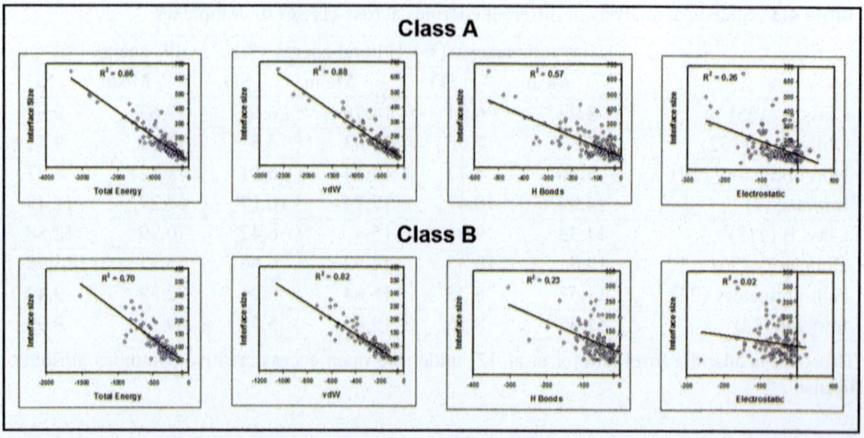

Fig. 4.12 The relationship between energy and its components (vdW, H bonds, electrostatics) with interface size among class A and class B complexes is shown (This image is adapted from Nilofer et al. [2] under the open access creative commons attribution license)

Table 4.2 Statistical analysis of different energies across different functional groups

	Electrostatic energy (%)		H bond energy (%)		vdW energy (%)	
	Mean	SD	Mean	SD	Mean	SD
All complexes (278)	11.26	8.7	15.03	6.54	74.93	11.37
Obligatory (208)	11.48	9.09	14.72	6.55	75.28	11.75
Enzyme (40)	9.73	9.41	15.49	5.03	74.78	9.64
Regulators (144)	11.87	9.28	14.72	7.22	75.07	12.47
Biological assembly (24)	9.20	9.04	13.41	3.98	77.39	10.6
Immune (18)	12.16	8.82	19.07	6.46	68.77	10.78
Non-obligatory (52)	10.08	7.01	14.89	6.16	75.68	9.42
Enzyme-inhibitors (25)	9.81	7.94	15.35	6.05	74.84	10.23
Regulator-inhibitors (27)	9.94	6.29	14.47	6.34	76.45	8.73

This data is adapted from Nilofer et al. [2] under the open access creative commons attribution license

4.11 Electrostatics, Interface Size, and Function

The effect of electrostatics in protein-protein binding is not insignificant (Tables 4.2 and 4.3). There is about 11.3% ± 8.7% contribution by electrostatic energy among protein-protein interfaces (Table 4.2). It increases with interface size specifically among non-obligatory regulator-inhibitors ($r^2 = 0.44$ with $p < 0.01$) as given in Table 4.4. Thus, there is a relation between interface electrostatics and regulator specific inhibitors. However, interfaces (class B) having less nonpolar residues than surface show 1.5-fold more electrostatic energy on average (Table 4.5). Electrostatic energy (8.8% ± 4.77% and $r^2 = 0.63$ with $p < 0.01$) among non-obligatory complexes within interfaces (class A) having more nonpolar residues than surface is

Table 4.3 Statistical analysis of different energies across classes of complexes

	Electrostatic energy (%)		H bond energy (%)		vdW energy (%)	
	Mean	SD	Mean	SD	Mean	SD
Class A (165)	8.99	6.84	14.78	6.62	77.63	9.6
Obligatory (132)	8.83	7.06	14.83	6.5	77.84	9.41
Non-obligatory (29)	8.76	4.77	14.14	6.81	78.12	8.77
Immune (4)	14.99	10.47	17.74	10.17	67.27	17.49
Class B (113)	14.33	9.96	15.4	6.42	70.99	12.58
Obligatory (76)	15.8	10.35	14.52	6.66	70.83	13.95
Non-obligatory (23)	11.57	8.78	15.84	5.21	72.59	9.49
Immune (14)	11.35	8.56	19.45	5.48	69.2	8.97

This data is adapted from Nilofer et al. [2] under the open access creative commons attribution license

Table 4.4 List of r^2 values between interface size and energy among complexes of known function

	Total interface energy		Van der Waals energy		H bond energy		Electrostatic energy	
	r^2	*p*-value	r^2	*p*-value	r^2	*p*-value	r^2	*p*-value
All complexes (278)	0.84	<0.01	0.87	<0.01	0.53	<0.01	0.15	<0.01
Obligatory (208)	0.85	<0.01	0.87	<0.01	0.57	<0.01	0.13	<0.01
Enzyme (40)	0.86	<0.01	0.86	<0.01	0.61	<0.01	0.19	0.01
Regulators (144)	0.82	<0.01	0.86	<0.01	0.5	<0.01	0.08	<0.01
Biological assembly (24)	0.79	<0.01	0.89	<0.01	0.54	<0.01	0	>0.01
Immune (18)	0.74	<0.01	0.67	<0.01	0.43	<0.01	0.21	>0.01
Non-obligatory (52)	0.76	<0.01	0.82	<0.01	0.24	<0.01	0.35	<0.01
Enzyme-inhibitors (25)	0.67	<0.01	0.75	<0.01	0.35	<0.01	0.18	>0.01
Regulator-inhibitors (27)	0.81	<0.01	0.85	<0.01	0.24	0.01	0.44	<0.01

This data is adapted from Nilofer et al. (2017) [2] under the open access creative commons attribution license

influencing our inference (Table 4.5). In an example complex, an obligatory regulatory DNA-binding protein (PDB: 3THO) interface is made of vdW (~92%) and electrostatics (~8%) without hydrogen bonds (Table 4.6). The role played by electrostatics for specificity is critical in this case.

Table 4.5 List of r^2 values between interface size and energy among class A (interface polarity is less than surface yet more than core) and class B (interface polarity more than surface and core) protein complexes

	Total interface energy		Van der Waals energy		H bond energy		Electrostatic energy	
	r^2	*p*-value	r^2	*p*-value	r^2	*p*-value	r^2	*p*-value
Class A (165)	0.86	<0.01	0.88	<0.01	0.57	<0.01	0.26	<0.01
Obligatory (132)	0.87	<0.01	0.88	<0.01	0.61	<0.01	0.24	<0.01
Non-obligatory (29)	0.81	<0.01	0.86	<0.01	0.2	0.01	0.63	<0.01
Immune (4)	0.85	>0.01	0.72	>0.01	0.76	>0.01	0.02	>0.01
Class B (113)	0.7	<0.01	0.82	<0.01	0.23	<0.01	0.02	>0.01
Obligatory (76)	0.69	<0.01	0.81	<0.01	0.24	<0.01	0.02	>0.01
Non-obligatory (23)	0.7	<0.01	0.83	<0.01	0.3	<0.01	0.02	>0.01
Immune (14)	0.77	<0.01	0.77	<0.01	0.38	>0.01	0.28	>0.01

This data is adapted from Nilofer et al. [2] under the open access creative commons attribution license

Table 4.6 Examples of interesting interfaces

PDB ID	Function	Name	Electrostatic energy (%)	H bonds (%)	vdW energy (%)
3B0Z	Obligatory regulatory	Protein transport	0	27.72	72.28
3THO	Obligatory regulatory	DNA-binding protein	8.28	0	91.72

This data is adapted from Nilofer et al. [2] under the open access creative commons attribution license

4.12 Conclusion

Protein-protein interfaces are characteristics of its binding features such as interface size, interface area, energy, and its components such as vdW, hydrogen bonds, and electrostatics. Each protein-protein interface is characterized by vdW, H bonds, and electrostatics. The protein-protein interfaces in protein complexes (heterodimers) are dominant in van der Waals (vdW) with selective H bonds and/or electrostatics with respect to specific molecular function.

Exercises

1. Illustrate the relationship between interface size and interface area.
2. Show the distribution of interface size in protein-protein complexes.
3. Illustrate the relationship between interface size and energy irrespective of molecular function.
4. Illustrate the relationship between interface size and vdW irrespective of molecular function.
5. Illustrate the relationship between interface size and H bonds.
6. Illustrate the relationship between interface size and electrostatics.
7. Enumerate interface H bonds and molecular function.
8. Enumerate interface electrostatics and molecular function.
9. Give an account of nonpolar interfaces using inter-residue analysis of complexes.
10. Give an account of polar interfaces using inter-residue analysis of complexes.

References

1. Sowmya G, Anita S, Kangueane P. Insights from the structural analysis of protein heterodimer interfaces. Bioinformation 2011 May 7;6(4): 137–143. PubMed PMID: 21572879; PubMed Central PMCID: PMC3092946.
2. Nilofer C, Sukhwal A, Mohanapriya A, Kangueane P. Protein-protein interfaces are vdW dominant with selective H-bonds and (or) electrostatics towards broad functional specificity. Bioinformation. 2017;13(6):164–73.
3. Sowmya G, Ranganathan S. Discrete structural features among interface residue-level classes. BMC Bioinform. 2015;16(Suppl 18):S8.
4. Sowmya G, Breen EJ, Ranganathan S. Linking structural features of protein complexes and biological function. Protein Sci. 2015;24(9):1486–94.

Homo Multimer Protein Complexes

5

5.1 Creation of Homo Multimer Protein Complex Dataset

We created a high-resolution (<3 Å) non-redundant (30% sequence similarity cut-off) homo multimer (trimer, tetramer, pentamer, hexamer, etc.) protein complex dataset from the protein data bank (PDB) as shown in Fig. 5.1. A total of 1267 complexes were thus identified as given in Table 5.1.

5.2 Distribution of Homo Multimer Protein Complexes

A distribution of known homo multimer protein complex structures at the protein data bank (PDB) is shown in Fig. 5.2. Homo tetramer is most represented in the dataset. This is interesting. The representation by tetramer protein complexes is implicit to the understanding of biological phenomenon. It is interesting to note that 3 mer and 6 mer are multiples. It should also be noted that 2 mer, 4 mer, 6 mer, and 8 mer are multiples. A comprehensive understanding of these complexes will shed insight to mechanism of protein complex formation.

5.3 Homo Trimer Protein Complexes

There are 324 known trimer protein complexes as shown in Fig. 5.1. This is an interesting observation. Therefore, it is of interest to study its geometric features at their interfaces in these complexes. Structure of a homo trimer glycosidase (1AM7) from enterobacteria phage lambda is shown as an example in Fig. 5.3. This complex is made of three homologous glycosidase monomers to form a functional lysozyme.

P. Kangueane, C. Nilofer, *Protein-Protein and Domain-Domain Interactions*,
https://doi.org/10.1007/978-981-10-7347-2_5

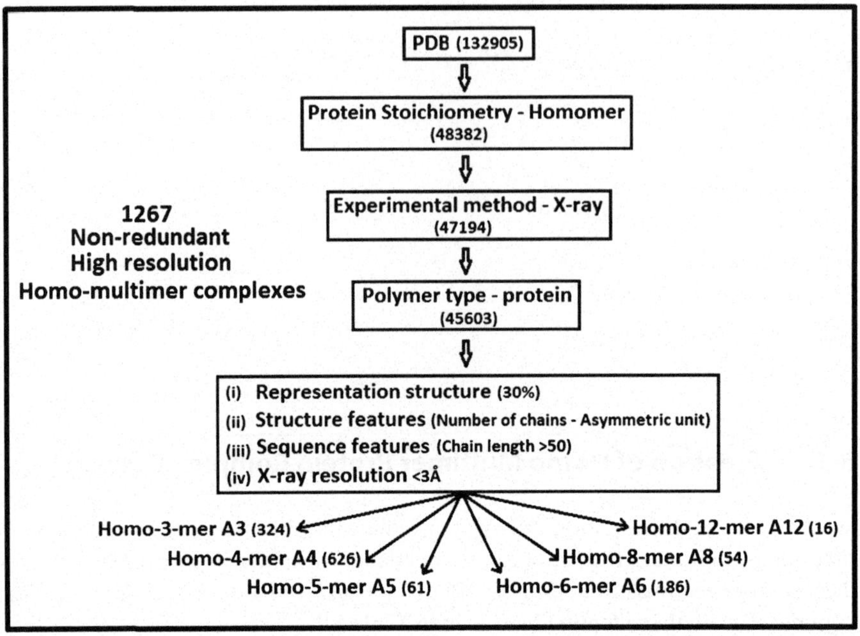

Fig. 5.1 A flowchart for the creation of a high-resolution (<3 Å) non-redundant (30% sequence similarity cutoff) homo multimer protein complex structure from protein data bank (PDB) [1] is shown

5.4 Homo Tetramer Protein Complexes

There are 626 known homo tetramer protein complexes as given in Fig. 5.1. These complexes are made of four homologous protein subunits. Structure of a homo tetramer endonuclease (1A79) from *Methanocaldococcus jannaschii* is shown in Fig. 5.4.

5.5 Homo Pentamer Protein Complexes

There are 61 known homo pentamer protein complexes as given in Fig. 5.1. Structure of a homo pentamer sugar-binding protein (4AVS) from *Homo sapiens* is shown in Fig. 5.5. The sugar binding protein complex is made of five serum amyloid p-component subunits.

Table 5.1 List of homo multimer protein complexes

Protein type	PDB-IDs
Homo-3-mer A3 (324)	1A12, 1AHS, 1AM7, 1AVQ, 1AVY, 1B5T, 1B77, 1BPO, 1BTG, 1C5E, 1C9K, 1CFM, 1CUN, 1D5F, 1DD1, 1EK9, 1EL6, 1EP5, 1FC3, 1GBN, 1HLQ, 1HX6, 1I8D, 1ICJ, 1IDP, 1J8F, 1JCD, 1JI7, 1JS0, 1JXZ, 1K4M, 1KKE, 1KNC, 1L5A, 1LUA, 1MWW, 1N3B, 1NAY, 1NKV, 1NMT, 1NXJ, 1O7K, 1O91, 1OMS, 1OTG, 1OXW, 1P32, 1PM4, 1PPR, 1PWB, 1QBZ, 1QDN, 1QP1, 1R6Z, 1RFX, 1RGX, 1RHG, 1RQ0, 1RV1, 1S55, 1S7O, 1SED, 1SG4, 1SI9, 1STZ, 1TUI, 1TZ0, 1U2M, 1UYJ, 1V8D, 1VKJ, 1VL0, 1W9Z, 1WDJ, 1WRV, 1WSP, 1WT6, 1WUD, 1XHO, 1Y4M, 1Y5E, 1YB0, 1YBF, 1YDM, 1YF9, 1YLF, 1YNB, 1YSR, 1YX1, 1ZCK, 1ZPV, 1ZRU, 2A5Z, 2ARH, 2B4I, 2B9B, 2BA2, 2BCM, 2BHW, 2BT9, 2C0A, 2CHC, 2CXI, 2CZ4, 2DJ6, 2E2A, 2EA7, 2EBO, 2EKM, 2F0C, 2FB5, 2FEF, 2FVH, 2G6Z, 2GFQ, 2GVH, 2HI9, 2I15, 2I9D, 2IEQ, 2IG8, 2IGT, 2INU, 2IUM, 2J5Z, 2JH2, 2MPR, 2O3C, 2O66, 2OTM, 2P4S, 2P90, 2Q01, 2Q0T, 2Q87, 2R3U, 2RE9, 2RGQ, 2V2H, 2VES, 2VHC, 2VQA, 2VRS, 2WN3, 2WQ4, 2WW2, 2WX3, 2X2R, 2X3H, 2XC1, 2XGF, 2XU8, 2YKO, 2YNY, 2YNZ, 2YZJ, 2Z7E, 3BYB, 3C6V, 3CM1, 3CPX, 3CQO, 3D8L, 3D9X, 3DA0, 3DAH, 3DC7, 3DLI, 3EEV, 3EMF, 3FD9, 3FK3, 3FLH, 3FRO, 3FUY, 3FX6, 3GDC, 3GGE, 3GKE, 3GQH, 3H0U, 3H35, 3H6X, 3HD5, 3HIA, 3HTN, 3I3F, 3IAC, 3IRS, 3JQY, 3K4I, 3KTO, 3L7G, 3LG7, 3LGI, 3LKE, 3MOG, 3MWP, 3N27, 3NKE, 3NTN, 3OGF, 3PR7, 3QK3, 3RPX, 3RWN, 3S6X, 3SAM, 3SOZ, 3SWF, 3T71, 3TAS, 3TK8, 3TY1, 3UBC, 3UD2, 3VBJ, 3VOP, 3WJ7, 3WV7, 3ZJB, 3ZXP, 4A0T, 4ADI, 4AE2, 4AGS, 4AIP, 4C47, 4C7R, 4CE7, 4CGC, 4E9X, 4EZC, 4F23, 4FAY, 4FN7, 4GIM, 4GOM, 4GQN, 4GVQ, 4H41, 4H9D, 4HI1, 4I6V, 4IWG, 4JDN, 4JJ2, 4L7D, 4LG1, 4LGO, 4LIJ, 4LIY, 4LYH, 4MEJ, 4MT4, 4NT9, 4OGG, 4OJ5, 4OK7, 4OOP, 4PBP, 4QKW, 4QTN, 4RFU, 4RMX, 4RU5, 4TME, 4U5R, 4UE0, 4UOF, 4USI, 4USX, 4UW7, 4UXE, 4W1T, 4WG2, 4WIA, 4WIP, 4WMF, 4X3N, 4XC5, 4XQA, 4Y7S, 4ZA1, 4ZCN, 4ZFX, 4ZM5, 5A95, 5APU, 5AZP, 5B2E, 5BMO, 5D4N, 5DMY, 5DO8, 5DWY, 5E57, 5ED9, 5EFV, 5EUR, 5FUS, 5G2F, 5G47, 5GP4, 5H3L, 5HK1, 5HT7, 5I2M, 5IP1, 5JBX, 5JPH, 5JQM, 5JS4, 5K21, 5MSX, 5N2C, 5NXF, 5P8W, 5T9Y, 5TE8, 5TOH, 5UCQ, 5V0Z, 5V13
Homo-4-merA4 (626)	1A79, 1AGQ, 1B25, 1B79, 1BAZ, 1BDF, 1BP5, 1BV4, 1BVI, 1C4P, 1CBM, 1CG2, 1CJX, 1D1J, 1D1Z, 1D2E, 1D8W, 1DIR, 1F2L, 1FJ0, 1FNU, 1FTR, 1FX3, 1G1Q, 1GPM, 1HJX, 1INL, 1J1J, 1J2R, 1J3W, 1JL2, 1JQQ, 1JS2, 1JTH, 1JUH, 1JYL, 1K6Y, 1L2G, 1L3A, 1LDD, 1M1L, 1M6S, 1MPX, 1N1Q, 1NBA, 1NC7, 1NI6, 1NJK, 1NYT, 1O58, 1O7J, 1O9Y, 1OA8, 1OFT, 1ON0, 1ORR, 1ORV, 1OU0, 1OVO, 1PJ3, 1PV1, 1Q15, 1Q5Y, 1QSM, 1QVC, 1QYN, 1RFZ, 1RLI, 1RTW, 1RXQ, 1RYI, 1S5A, 1SCE, 1SU1, 1SUW, 1T3U, 1TBU, 1TJV, 1TK9, 1TLF, 1TVX, 1TZJ, 1U3A, 1U4F, 1U69, 1U8V, 1U9Y, 1UB3, 1UDR, 1UNK, 1V8Q, 1V9D, 1VDD, 1VE5, 1VH1, 1VIM, 1VJ0, 1VK8, 1VKN, 1VL6, 1VQ3, 1VR6, 1W6U, 1WOC, 1WPG, 1WQS, 1WS8, 1WTY, 1WWH, 1WWJ, 1WZO, 1X0G, 1XKL, 1XM3, 1XPJ, 1XPP, 1XQ4, 1XTT, 1XWR, 1Y1L, 1Y1O, 1Y2M, 1Y66, 1Y82, 1Y9I, 1YAX, 1YBK, 1YD6, 1YJ7, 1YLL, 1YNR, 1YP2, 1YVW, 1YXM, 1Z0S, 1ZQ7, 1ZTC, 1ZWY, 2A2U, 2A5H, 2A5V, 2A9G, 2AE6, 2AFG, 2AHU, 2AZ5, 2B3Z, 2C46, 2CDC, 2CZC, 2D16, 2D32, 2D3E, 2D3W, 2DFT, 2DGE, 2DSY, 2DVT, 2E7P, 2ELC, 2EPI, 2ESN, 2FA8, 2FPD, 2FR5, 2FZV, 2GF6, 2GJD, 2GM5, 2GTA, 2GZ3, 2H5N, 2HIY, 2HJM, 2HLZ, 2HSJ, 2HX1, 2I9F, 2IAZ, 2IBO, 2ID5, 2IEC, 2IHT, 2IS5, 2IVO, 2IXC, 2J91, 2JGS, 2JLP,

(continued)

Table 5.1 (continued)

Protein type	PDB-IDs
	2NLZ, 2NTK, 2NYT, 2NZC, 2O18, 2OEQ, 2OGF, 2OIE, 2OIW, 2OLK, 2OQG, 2OX7, 2OZK, 2P2D, 2P3N, 2P63, 2P9X, 2PBY, 2PHP, 2PI8, 2PK2, 2PLI, 2POC, 2PQQ, 2PZH, 2Q73, 2Q81, 2QAP, 2QV6, 2QW6, 2QWZ, 2R0H, 2R3Z, 2R4F, 2RFQ, 2V66, 2V7G, 2VHS, 2VL9, 2VP4, 2VUN, 2VV8, 2VXW, 2VZY, 2W4I, 2WKC, 2WKK, 2WSH, 2WUL, 2X4H, 2X5D, 2X7J, 2XBL, 2XHZ, 2XIG, 2Y32, 2Y7J, 2YWM, 2YZH, 2Z3H, 2Z67, 2Z8U, 2ZGI, 2ZIH, 2ZRU, 3AAI, 3ABZ, 3AEK, 3AK2, 3ALU, 3AQQ, 3AUP, 3B8F, 3BB6, 3BF0, 3BG3, 3BH1, 3BJU, 3BP1, 3BPZ, 3BU2, 3BU8, 3C1M, 3C2U, 3C3N, 3C4R, 3C6F, 3CE9, 3CI7, 3CMB, 3CNV, 3CO2, 3CQ2, 3CSW, 3CVE, 3CXJ, 3CZQ, 3D55, 3D8K, 3DDV, 3DH4, 3DLU, 3DTY, 3DZA, 3E8M, 3E8P, 3EDC, 3EDT, 3EGI, 3ESI, 3ETN, 3EYT, 3FAU, 3FEF, 3FHK, 3FKA, 3FKF, 3FMC,3FPA, 3FPC, 3FXA, 3FZV, 3G2E, 3G5N, 3G5P, 3G7K, 3GBH, 3GEK, 3GG2, 3GGM, 3GH1, 3GOD, 3GWZ, 3GXI, 3GYG, 3H5N, 3H7A, 3HCG, 3HH0, 3HM0, 3HN3, 3HQ0, 3HVY, 3HWC, 3I3V, 3I7U, 3I9S, 3IB6, 3IC8, 3IG2, 3INN, 3IP1, 3ISR, 3IWV, 3IXQ, 3JUR, 3JZE, 3K17, 3K2G, 3K4T, 3K90, 3K9R, 3KLT, 3KVE, 3KW0, 3KZG, 3L0G, 3L27, 3L2H, 3L7I, 3LD9, 3LM4, 3LOU, 3LSZ, 3LUQ, 3M0M, 3M0Z, 3M9U, 3MDN, 3MKH, 3MQC, 3MSQ, 3MUP, 3MXQ, 3N07, 3N2W, 3N3Y, 3NB0, 3NG4, 3NJA, 3NKZ, 3NND, 3NO6, 3NPG, 3NWH, 3NWZ, 3NXZ, 3O94, 3O9Z, 3OJC, 3ON7, 3ONQ, 3OQL, 3OVK, 3PAY, 3PFN, 3PIB, 3PN9, 3PVJ, 3PVS, 3QVI, 3R2U, 3RAM, 3RD4, 3RMT, 3RO0, 3RQA, 3RXZ, 3S2Y, 3S6G, 3SE5, 3SHO, 3SWO, 3THR, 3TIA, 3TIK, 3TR5, 3TR9, 3TSI, 3TUO, 3U02, 3U35, 3UG7, 3UO9, 3UUW, 3VIR, 3VNI, 3VPC, 3VR0, 3VSV, 3W68, 3WB0, 3WCC, 3ZL9, 3ZZH, 4A46, 4A57, 4A64, 4A6H, 4A9Z, 4ADM, 4AGR, 4AUM, 4AVL, 4B2O, 4BL6, 4BV1, 4BWW, 4CCK, 4CE8, 4CQ4, 4CY8, 4D1N, 4DPL, 4DSY, 4EWT, 4F0Q, 4FM3, 4FO8, 4G2N, 4GHK, 4GQY, 4GX0, 4HAD, 4HGN, 4HH2, 4HRF, 4IAB, 4IF4, 4IPY, 4IT5, 4IVE, 4J4G, 4J4Y, 4JEA, 4JF7, 4JIB, 4K5U, 4KD5, 4KGM, 4KOP, 4KSA, 4KTB, 4L4Q, 4L5J, 4L8F, 4LGV, 4LOC, 4LRL, 4LY4, 4M1X, 4M20, 4M3L, 4M3P, 4M6R, 4MER, 4MIY, 4MKQ, 4MS2, 4MY5, 4N8G, 4N8R, 4NB5, 4NZG, 4O4A, 4O59, 4OC8, 4OHF, 4OJU, 4OM3, 4OM5, 4OO2, 4OUH, 4OX0, 4P3G, 4P7D, 4PCA, 4PLS, 4PSM, 4PSN, 4PTZ, 4PWA, 4Q1T, 4Q2J, 4QFW, 4QHB, 4QHZ, 4QNN, 4QPO, 4R0C, 4R7K, 4R8X, 4S1V, 4TKD, 4TW5, 4TYB, 4U1R, 4U6B, 4U8P, 4UM7, 4UOX, 4UXD, 4W7Z, 4XCB, 4XCH, 4XUU, 4Y6Q, 4YEG, 4YLB, 4YWA, 4Z6M, 4ZBO, 4ZHT, 4ZO4, 4ZTB, 5BQ2, 5BQP, 5BZB, 5C2I, 5C8A, 5C9F, 5CUY, 5D05, 5D60, 5DEC, 5DH1, 5DLE, 5DNC, 5E9G, 5EGF, 5EGQ, 5ELP, 5FB7, 5FC9, 5FGS, 5FJD, 5FTA, 5G0A, 5GLJ, 5H1A, 5HK7, 5HWN, 5IBZ, 5IE9, 5IH3, 5IN4, 5IVG, 5J78, 5J7R, 5J87, 5J8R, 5JA3, 5JBS, 5JFN, 5JU6, 5K90, 5KR2, 5LC9, 5LCY, 5MPW, 5MU4, 5MV0, 5N2D, 5O4J, 5SUY, 5T07, 5T0Z, 5TGN, 5TIA, 5TLC, 5TTX, 5TVL, 5U55, 5UBU, 5UDF, 5UNC, 5UPB, 5V84, 5VOH, 5W15, 5XAC, 5XDC, 5XTA, 6Q21
Homo-5-merA5 (61)	1C4Q, 1IAS, 1JG5, 1NLQ, 1O66, 1S7G, 1T0T, 1VR4, 1XMI, 1Y2I, 1Y60, 2B98, 2C92, 2GSC, 2GUV, 2HEX, 2PGC, 2RLD, 2VQZ, 2WN9, 3BY7, 3CK6, 3DCX, 3DRZ, 3DTZ, 3KIH, 3OAE, 3P05, 3QKB, 3SPU, 3SXP, 3TDS, 3WZ2, 4AVS, 4DMI, 4DWL, 4EXW, 4FP5, 4HFI, 4J2N, 4J4V, 4MOZ, 4N8M, 4OM4, 4PZN, 4RHS, 4WBA, 4WD8, 5A12, 5BXD, 5H5T, 5HLR, 5INR, 5IQQ, 5JBL, 5JIE, 5L37, 5LZG, 5MOG, 5N78, 5XPC

(continued)

Table 5.1 (continued)

Protein type	PDB-IDs
Homo-6-merA6 (186)	1A2V, 1ATJ, 1B4A, 1BQ7, 1BU1, 1DM5, 1E9R, 1F6R, 1GUT, 1IQP, 1JX7, 1KNX, 1KQN, 1LVU, 1NE7, 1NMO, 1NSL, 1O9I, 1T8S, 1TT7, 1U7K, 1V7Z, 1VKE, 1WEK, 1WMG, 1XK8, 1XX7, 1YKB, 1ZCC, 1ZEI, 2A1F, 2CE7, 2CEV, 2COV, 2D00, 2DDZ, 2DQB, 2DV6, 2EKD, 2F2H, 2G3M, 2GJV, 2GK3, 2GLL, 2GMY, 2H12, 2H7C, 2HV2, 2IUU, 2J6P, 2J9L, 2JF9, 2JLM, 2OBA, 2OGJ, 2P10, 2PMU, 2Q3S, 2Q80, 2QEZ, 2R5U, 2V1O, 2VHX, 2VXH, 2W4L, 2W5E, 2XF7, 2XMV, 2Y6E, 2Y75, 2ZBL, 2ZDS, 2ZFZ, 3A11, 3A1M, 3B59, 3BBB, 3BEY, 3BJK, 3BMA, 3BV6, 3BZW, 3D03, 3D19, 3D7J, 3E0O, 3EJX, 3FD8, 3GWJ, 3H8G, 3HBX, 3HE1, 3IG4, 3IGQ, 3IH6, 3IS5, 3ISA, 3KOP, 3KVP, 3L28, 3L7P, 3LYU, 3M1R, 3MBH, 3MQ1, 3MW9, 3N2N, 3NWY, 3OVG, 3OZB, 3PW3, 3SBA, 3SLG, 3T3W, 3T94, 3VZD, 3WCM, 3WSG, 3WU4, 4BLP, 4BLQ, 4C01, 4C9S, 4D2E, 4DZ4, 4FLN, 4FZQ, 4G2S, 4GM6, 4H83, 4HGZ, 4HUD, 4I83, 4L7Q, 4L80, 4LK3, 4LKB, 4LZK, 4MM1, 4MY0, 4N2X, 4NTK, 4OFC, 4ONX, 4P7T, 4PH0, 4PPH, 4PZO, 4R1Q, 4R2X, 4RHW, 4RHW, 4RVP, 4UDG, 4URQ, 4V1X, 4WKG, 4X9C, 4XGU, 4XLZ, 4YB6, 4ZDA, 4ZLA, 5A4U, 5AA6, 5CAI, 5DM5, 5F4H, 5FB3, 5G3P, 5GMZ, 5GQ1, 5GUD, 5H0T, 5HJX, 5IAO, 5IT5, 5JDB, 5JU5, 5KKO, 5NAV, 5SWC, 5TSH, 5TUW, 5U9C, 5WMF
Homo-8-mer A8 (54)	1B9L, 1E2T, 1EKJ, 1KCA, 1N0E, 1QK1, 1T35, 1T5R, 1UF3, 1Z4A, 2AEN, 2ANE, 2CFX, 2FY8, 2IB5, 2IGS, 2IZY, 2J58, 2NR5, 2OZR, 2P9J, 2Q0S, 2QMI, 2R8E, 2V9U, 2VUB, 2ZBC, 2ZNZ, 3AJ1, 3B8O, 3BJW, 3ECS, 3FIJ, 3FXG, 3GR1, 3HZN, 3MKV, 3OOJ, 3OQB, 3RST, 4IAJ, 4JHM, 4KWB, 4L0O, 4PHX, 4UNT, 4WP2, 5DZN, 5ECY, 5GJA, 5HSA, 5JQV, 5K8P, 5KAK
Homo-12-mer A12 (16)	1NGK, 1VDM, 1ZS3, 2CF7, 2E6G, 2EWC, 3AK8, 3DTD, 3FP9, 3MQ7, 3QJG, 3R90, 4CYB, 4MH2, 4RAP, 5T11

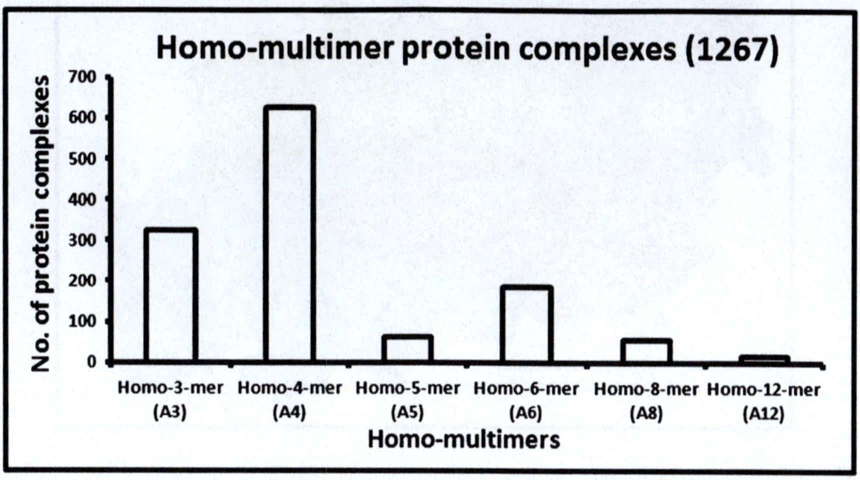

Fig. 5.2 Distribution of homo multimer protein complex structures at the protein data bank (PDB). Homo tetramer is most represented in the dataset

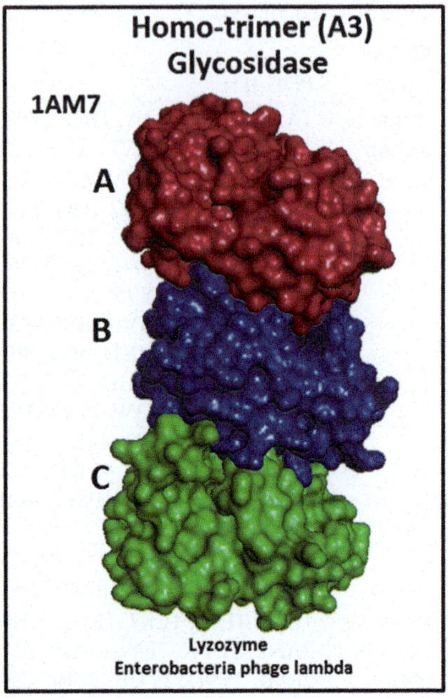

Fig. 5.3 Structure of a homo trimer glycosidase (1AM7) from enterobacteria phage lambda is drawn using Discovery Studio Visualizer™ version 16.1 [2]. This complex is made of three homologous glycosidase monomers to form a functional lysozyme. Homo multimer protein binding is critical in the formation of a functional complex

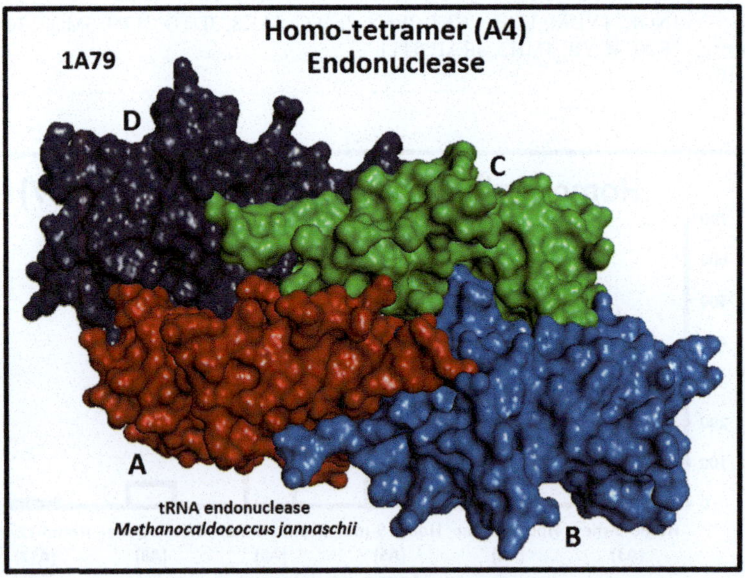

Fig. 5.4 Structure of a homo tetramer endonuclease (1A79) from *Methanocaldococcus jannaschii* is drawn using Discovery Studio Visualizer™ version 16.1 [2]. This complex is made of four homologous protein subunits. Homo multimer protein binding is critical in the formation of a functional complex

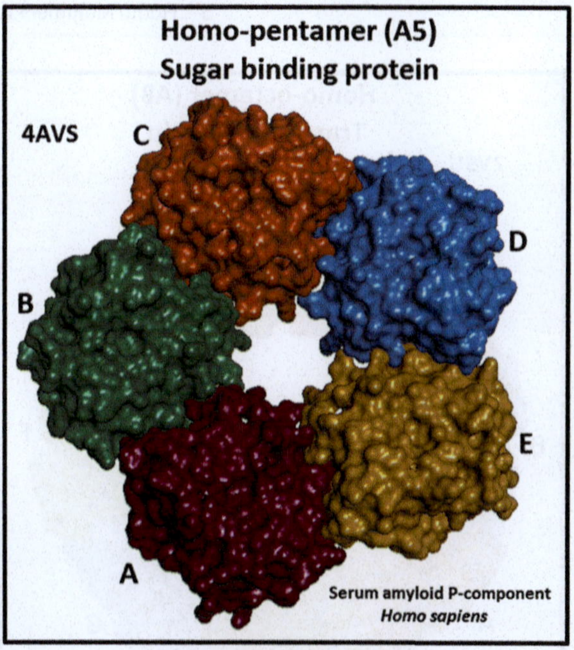

Fig. 5.5 Structure of a homo pentamer sugar-binding protein (4AVS) from *Homo sapiens* is drawn using Discovery Studio Visualizer™ version 16.1 [2]. The sugar-binding protein complex is made of five serum amyloid p-component subunits. Homo multimer protein binding is critical in the formation of a functional complex

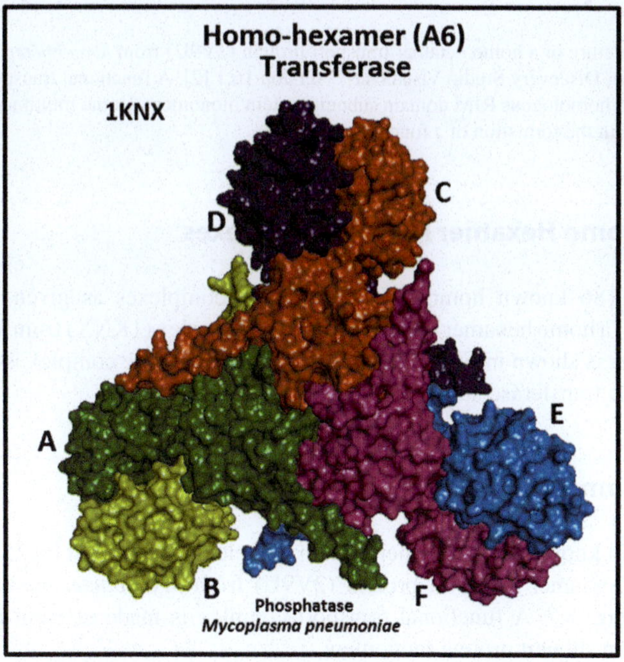

Fig. 5.6 Structure of a homo hexamer transferase protein complex (1KNX) from *Mycoplasma pneumoniae* is drawn using Discovery Studio Visualizer™ version 16.1 [2]. A functional phosphatase complex is made of six homologous transferase subunits. Homo multimer protein binding is critical in the formation of a functional complex

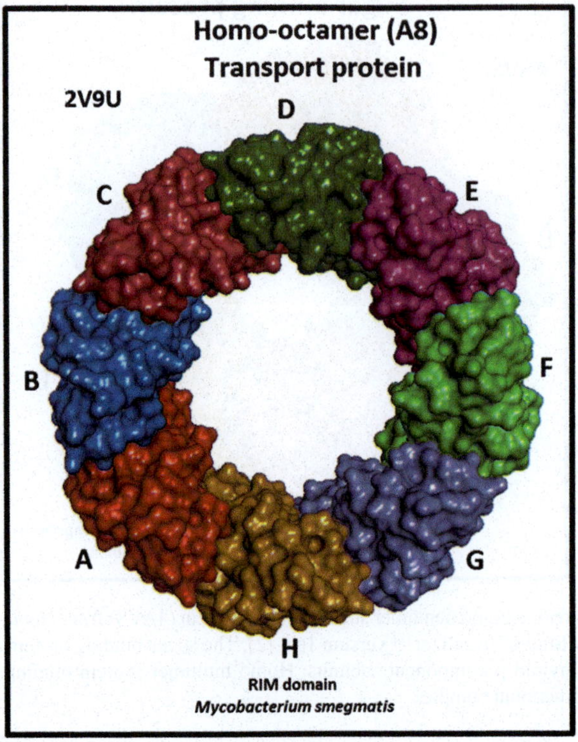

Fig. 5.7 Structure of a homo octamer transport protein (2V9U) from *Mycobacterium smegmatis* is drawn using Discovery Studio Visualizer™ version 16.1 [2]. A functional transport complex is made of eight homologous RIM domain subunit protein monomers. Homo multimer protein binding is critical in the formation of a functional complex

5.6 Homo Hexamer Protein Complexes

There are 186 known homo hexamer protein complexes as given in Fig. 5.1. Structure of a homo hexamer transferase protein complex (1KNX) from *Mycoplasma pneumoniae* is shown in Fig. 5.6. A functional phosphatase complex is made of six homologous transferase subunits.

5.7 Homo Octamer Protein Complexes

There are 54 known homo octamer protein complexes as given in Fig. 5.1. Structure of a homo octamer transport protein (2V9U) from *Mycobacterium smegmatis* is shown in Fig. 5.7. A functional transport complex is made of eight homologous RIM domain subunit protein monomers.

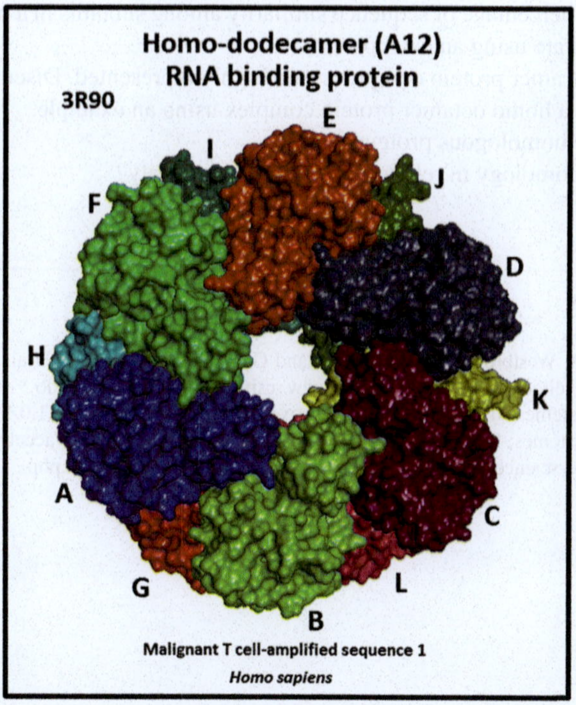

Fig. 5.8 Structure of a homo dodecamer RNA-binding protein complex (3R90) from *Homo sapiens* is drawn using Discovery Studio Visualizer™ version 16.1 [2]. This RNA-binding protein complex is made of 12 homologous protein subunits. Homo multimer protein binding is critical in the formation of a functional complex

5.8 Homo Dodecamer Protein Complexes

There are 16 known homo dodecamer protein complexes as given in Fig. 5.1. Structure of a homo dodecamer RNA-binding protein complex (3R90) from *Homo sapiens* is shown in Fig. 5.8. This RNA-binding protein complex is made of 12 homologous protein subunits.

Exercises

1. What are homo multimer protein complexes?
2. Illustrate a flowchart to create a dataset of homo multimer protein complexes.
3. Show the distribution of homo multimer protein complexes based on the number of subunits.
4. Illustrate a homo pentamer protein complex using an example.
5. How many subunits constitute a dodecamer and illustrate using an example?

6. State the percentage of sequence similarity among subunits in a homo multimer and illustrate using an example.
7. Homo tetramer protein complexes are highly represented. Discuss in detail.
8. Illustrate a homo octamer protein complex using an example.
9. What are homologous proteins?
10. What is homology in relation to sequence similarity?

References

1. Berman HM, Westbrook J, Feng Z, Gilliland G, Bhat TN, Weissig H et al. The protein data bank. 2000. [cited: 2017 Aug 1]. http://www.rcsb.org/pdb/home/home.do
2. Dassault Systemes BIOVIA, BIOVIA Discovery Studio Visualizer, v16.1.0.15350, San Diego: Dassault Systemes; 2015. [cited: 2017 Aug 1]. Available from: http://accelrys.com/products/collaborative-science/biovia-discovery-studio/visualization-download.php

Hetero Multimer Protein Complexes

6

6.1 Creation of Hetero Multimer Protein Complex Dataset

We created a high-resolution (<3 Å) non-redundant (30% sequence similarity cut-off) homo multimer (ABC, A2B, ABCD, A2B2, A2B2C2, A3B3) protein complex dataset from the protein data bank (PDB) [1] as shown in Fig. 6.1. A total of 384 complexes were thus identified as given in Table 6.1.

6.2 Distribution of Hetero Multimer Protein Complexes

A distribution of known hetero multimer protein complex structures at the protein data bank (PDB) is shown in Fig. 6.2. Hetero tetramer (A2B2) is most represented in the dataset. This is interesting. The representation by A2B2 tetramer protein complexes is implicit to the understanding of biological phenomenon. A comprehensive understanding of these complexes will shed insight to mechanism of multimer protein complex formation towards their role.

6.3 ABC Hetero Trimer Protein Complex

There are 81 ABC trimer protein complexes as shown in Fig. 6.1. This is an interesting dataset. Therefore, it is of interest to study its geometric features at their interfaces in these complexes. A comprehensive understanding of large number of ABC trimer protein complexes will provide valuable insights to trimer protein assembly and its role in biological system. Structure of an ABC hetero trimer urease protein complex from *Sporosarcina pasteurii* is shown in Fig. 6.3. Protein subunits A, B, and C are non-homologous and show sequence dissimilarity among themselves.

P. Kangueane, C. Nilofer, *Protein-Protein and Domain-Domain Interactions*,
https://doi.org/10.1007/978-981-10-7347-2_6

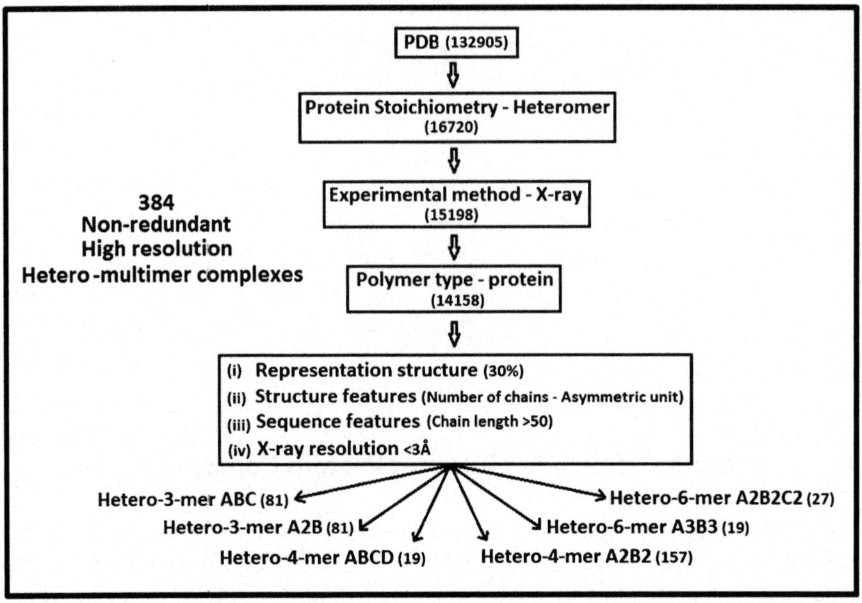

Fig. 6.1 A flowchart for the creation of a high-resolution (<3 Å) non-redundant (at 30% sequence similarity cutoff) hetero multimer protein complex structure from protein data bank (PDB) is shown

Table 6.1 List of hetero multimer protein complexes

Protein type	PDB-IDs
Hetero-3-merABC (81)	1IKN, 1M93, 1PBY, 1RK8, 1SR4, 1XKP, 1Y5I, 2ADV, 2AZE, 2DSR, 2HUE, 2IVF, 2J8C, 2NUT, 2P58, 2PBD, 2PU9, 2QFA, 2V8Q, 2VJ0, 2W2N, 2XQN, 2ZXE, 3A1J, 3AA6, 3BOW, 3BPS, 3BT1, 3BWU, 3CI0, 3FDS, 3HTK, 3IP4, 3IUR, 3N44, 3N9O, 3P5J, 3QKS, 3S8G, 3TVJ, 3U1J, 3U28, 3W15, 4A8X, 4CAY, 4CY2, 4EXT, 4F0Z, 4G91, 4H9N, 4HAT, 4HR6, 4ILG, 4NN5, 4OWT, 4PSW, 4Q5E, 4U1E, 4UBP, 4W4L, 4XPM, 5A53, 5AJA, 5ANR, 5AWW, 5CWW, 5F0J, 5F1B, 5FO8, 5G5G, 5H66, 5J1S, 5J3T, 5KDO, 5M5E, 5MC9, 5MJ3, 5T5W, 5UBP, 5VLL, 5WGB
Hetero-3-merA2B (81)	1A7C, 1BVY, 1C7C, 1DKG, 1EER, 1F6F, 1GHQ, 1HWG, 1IK9, 1JZD, 1L0O, 1NVU, 1OY3, 1OYV, 1SG1, 1U2G, 1X79, 1YA5, 1YK0, 2A19, 2A5Y, 2B2Y, 2BCN, 2CJS, 2E4M, 2EQ7, 2FM8, 2YVJ, 3D36, 3DPB, 3DWG, 3E95, 3ECH, 3EJJ, 3F1I, 3HPW, 3ISM, 3MHP, 3ONL, 3OWT, 3Q68, 3S2K, 3T12, 3T97, 3TKZ, 3UTM, 3ZOQ, 4AQR, 4D0K, 4DOS, 4EB7, 4HH3, 4HNJ, 4JMF, 4JO9, 4KT5, 4LN0, 4LXR, 4N1C, 4N3Y, 4N7V, 4OWF, 4PL8, 4UI0, 4W4O, 4ZJB, 5A6W, 5AER, 5B7I, 5EFW, 5FX8, 5H07, 5HUY, 5HVZ, 5HYC, 5IBW, 5LG4, 5M5R, 5VE9, 5WNW, 5XSD
Hetero-4-merABCD (19)	1QHH, 2GAG, 2J3T, 2NPS, 2P22, 2WJN, 3HRZ, 3VH5, 4FK5, 4GS7, 4I6M, 4UQI, 4WY4, 5BNX, 5DIS, 5HCC, 5JH5, 5MQU, 5TCS

(continued)

Table 6.1 (continued)

Protein type	PDB-IDs
Hetero-4-merA2B2 (157)	1B5F, 1B5F, 1CCW, 1CCW, 1CD9, 1DNU, 1DNU, 1E1H, 1E1H, 1EF1, 1EF1, 1EFU, 1F2U, 1F2U, 1FS1, 1FS1, 1GPQ, 1GPQ, 1GVN, 1GVN, 1I5K, 1I7Q, 1I7Q, 1JNR, 1JNR, 1K2X, 1K2X, 1KZY, 1M2Z, 1MOX, 1OAO, 1OAO, 1REW, 1REW, 1UNN, 1UNN, 1W6S, 1WVE, 1WVE, 2BO9, 2BO9, 2F9I, 2F9I, 2GH0, 2GH0, 2GIA, 2GIA, 2LTN, 2LTN, 2OIZ, 2Q4G, 2Q4G, 2QIY, 2QIY, 2QM6, 2QM6, 2RHK, 2RHK, 2VOF, 2VOF, 2VSX, 2XPI, 2XTS, 2Y8N, 2Y8N, 2YC4, 2ZYZ, 2ZYZ, 3AAW, 3BP8, 3D0G, 3D87, 3GS2, 3GS2, 3H6P, 3H6P, 3IT4, 3IT4, 3IWC, 3IWC, 3KDF, 3KDF, 3M91, 3M91, 3MM5, 3MM5, 3N9U, 3N9U, 3O1J, 3O6Q, 3P8M, 3RPF, 3RPF, 3T6Q, 3T6Q, 3U7Q, 3UCS, 3UCS, 3V64, 3VTI, 3ZV0, 4EKX, 4EKX, 4ELY, 4ELY, 4GSL, 4H2W, 4H2W, 4HRN, 4ICG, 4KEH, 4KEH, 4LM6, 4LM6, 4LRT, 4LRT, 4MCT, 4N7R, 4N88, 4OD9, 4OD9, 4PV2, 4PV2, 4QPJ, 4R3U, 4U5W, 4U5W, 4UYP, 4V02, 4WES, 4WRL, 4X28, 4X28, 4YB8, 4YB8, 4YMU, 4YTW, 5ADU, 5AHU, 5AOQ, 5CW4, 5D28, 5E9T, 5IAA, 5IAA, 5IFG, 5IU0, 5IU0, 5J43, 5JAA, 5MVW, 5MVW, 5MZW, 5MZW, 5NT7, 5UHK, 5UN5
Hetero-6-merA3B3 (19)	1K90, 1PYA, 1XU1, 2DE6, 2FYZ, 3EJ9, 3GCM, 3M6M, 3MKQ, 3OA8, 3TWI, 4FAS, 4FZ0, 4GIP, 4HEM, 4MMS, 4RJF, 5ENQ, 5WVC
Hetero-6-merA2B2C2 (27)	1EEX, 1FI8, 1JY2, 1L1O, 1MTY, 1N62, 1O94, 2BS2, 2OOX, 2VPZ, 3NVW, 3OR1, 3SJL, 3U52, 4A8J, 4AUQ, 4DJD, 4GDK, 4HI0, 4KNG, 5A0Y, 5BNV, 5C2V, 5EXE, 5KQ4, 5N6Y, 5TDS

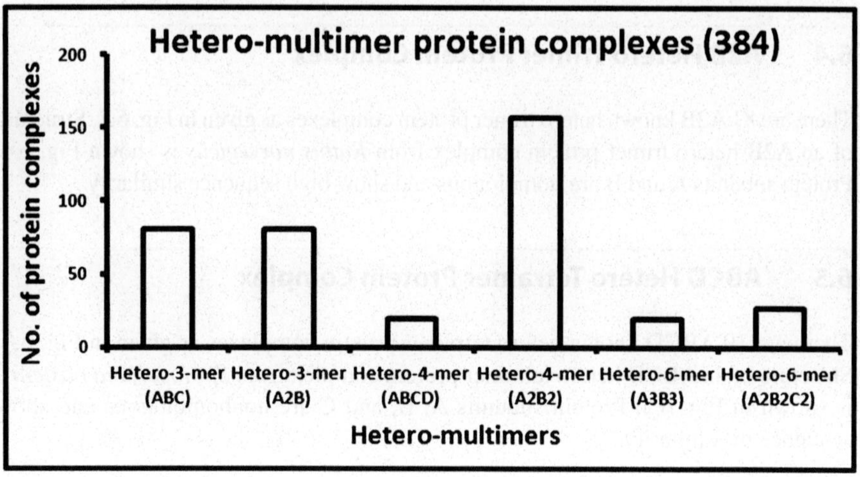

Fig. 6.2 Distribution of hetero multimer protein complex structures at the protein data bank (PDB). Hetero tetramer (A2B2) is most represented in the dataset. This is interesting

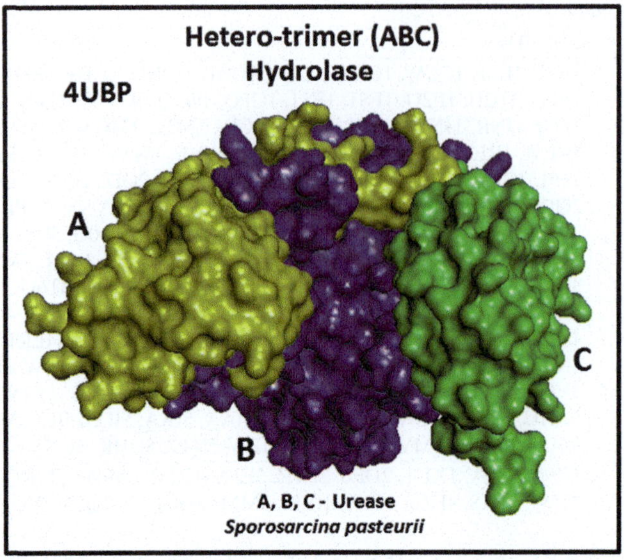

Fig. 6.3 Structure of an ABC hetero trimer urease protein complex from *Sporosarcina pasteurii* is drawn using Discovery Studio Visualizer™ version 16.1 [2]. Protein subunits A, B, and C are non-homologous and show sequence dissimilarity

6.4 A2B Hetero Trimer Protein Complex

There are 81 A2B known hetero trimer protein complexes as given in Fig. 6.1. Structure of an A2B hetero trimer protein complex from *Rattus norvegicus* is shown Fig. 6.4. Protein subunits A and B are homologous and show high sequence similarity.

6.5 ABCD Hetero Tetramer Protein Complex

There are 19 ABCD known hetero tetramer protein complexes as given in Fig. 6.1. Structure of an ABCD hetero tetramer protein complex from *Sporosarcina pasteurii* is shown in Fig. 6.5. Protein subunits A, B, and C are nonhomologous and show sequence dissimilarity.

6.6 A2B2 Hetero Tetramer Protein Complex

There are 157 A2B2 known hetero tetramer protein complexes as given in Fig. 6.1. Structure of an A2B2 hetero tetramer DNA-binding protein complex from *Homo sapiens* is shown in Fig. 6.6. There are two homologous and two nonhomologous subunits in this protein complex.

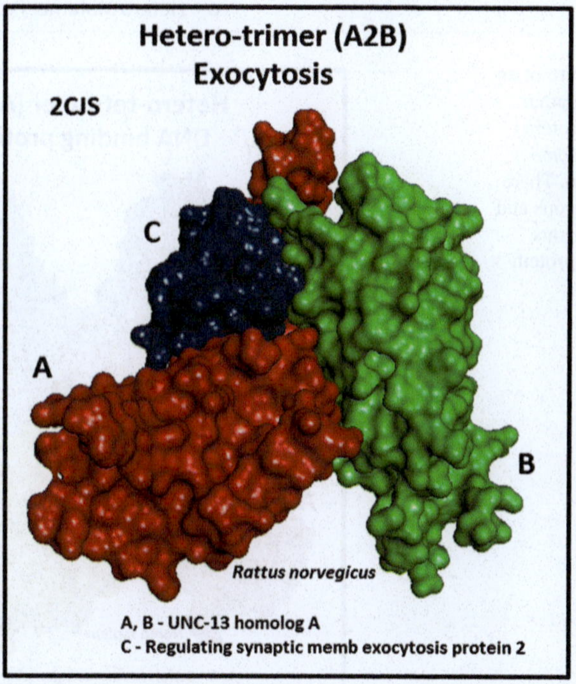

Fig. 6.4 Structure of an A2B hetero trimer protein complex from *Rattus norvegicus* is drawn using Discovery Studio Visualizer™ version 16.1 [2]. Protein subunits A and B are homologous and show high sequence similarity

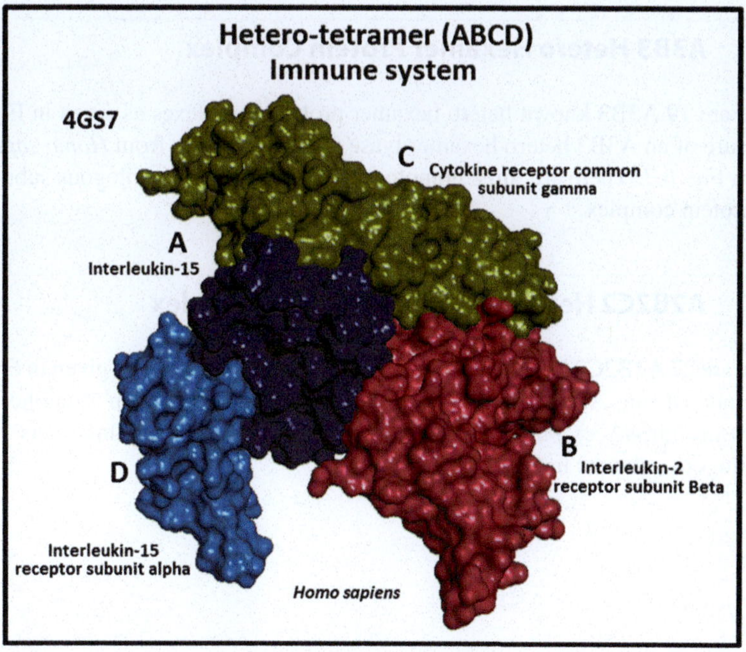

Fig. 6.5 Structure of an ABCD hetero tetramer protein complex from *Sporosarcina pasteurii* is drawn using Discovery Studio Visualizer™ version 16.1 [2]. Protein subunits A, B, and C are nonhomologous and show sequence dissimilarity

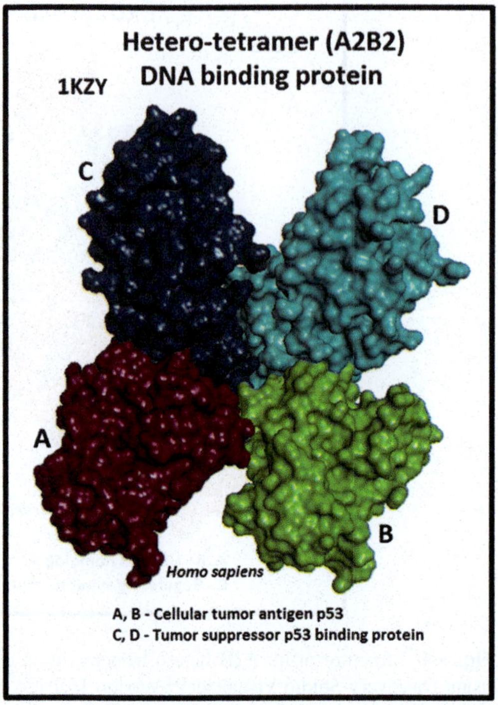

Fig. 6.6 Structure of an A2B2 hetero tetramer DNA-binding protein complex from *Homo sapiens* is shown. There are two homologous and two nonhomologous subunits in this protein complex

6.7 A3B3 Hetero Hexamer Protein Complex

There are 19 A3B3 known hetero hexamer protein complexes as given in Fig. 6.1. Structure of an A3B3 hetero hexamer lyase protein complex from *Homo sapiens* is shown Fig. 6.7. There are three homologous and three nonhomologous subunits in this protein complex.

6.8 A2B2C2 Hetero Hexamer Protein Complex

There are 27 A2B2C2 known hetero hexamer protein complexes as given in Fig. 6.1. Structure of an A2B2C2 hetero hexamer transferase protein complex from *Schizosaccharomyces pombe* is shown in Fig. 6.8. There are three sets of two homologous subunits in this protein complex.

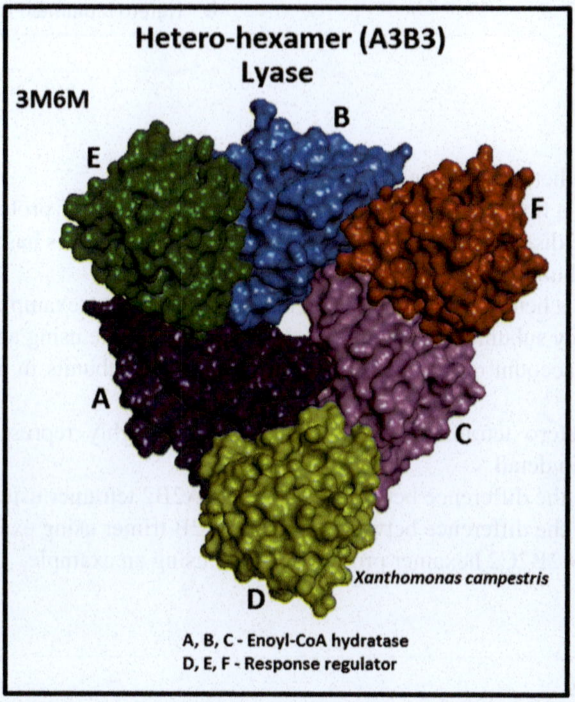

Fig. 6.7 Structure of an A3B3 hetero hexamer lyase protein complex from *Homo sapiens* is drawn using Discovery Studio Visualizer™ version 16.1 [2]. There are three homologous and three non-homologous subunits in this protein complex

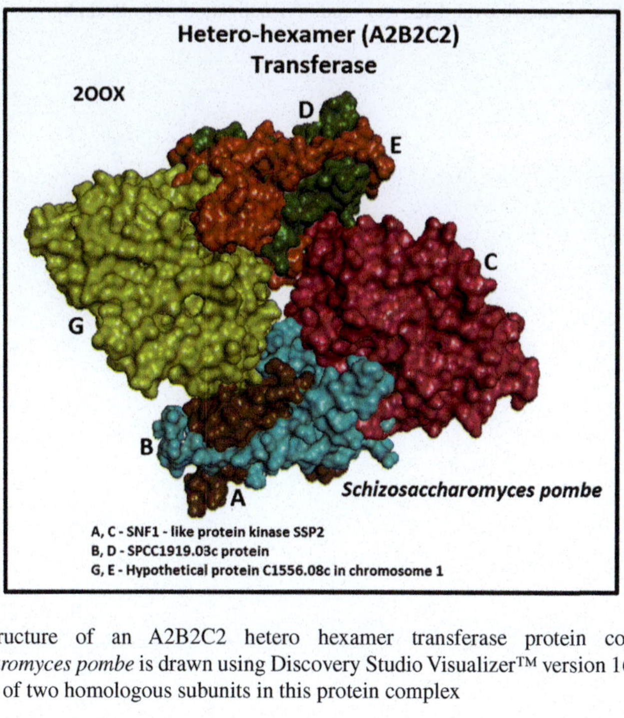

Fig. 6.8 Structure of an A2B2C2 hetero hexamer transferase protein complex from *Schizosaccharomyces pombe* is drawn using Discovery Studio Visualizer™ version 16.1 [2]. There are three sets of two homologous subunits in this protein complex

Exercises

1. What are hetero multimer protein complexes?
2. Illustrate a flowchart to create a dataset of hetero multimer protein complexes.
3. Show the distribution of hetero multimer protein complexes based on the number of subunits.
4. Illustrate a hetero ABC trimer protein complex using an example.
5. How many subunits constitute a pentamer and illustrate using an example?
6. Give an account of sequence dissimilarity among subunits in an ABC hetero trimer.
7. A2B2 hetero tetramer protein complexes are highly represented in PDB. Discuss in detail.
8. Illustrate the difference between ABCD and A2B2 tetramer using examples.
9. Illustrate the difference between ABC and A2B trimer using examples.
10. Discuss A2B2C2 hexamer protein complex using an example.

References

1. Berman HM, Westbrook J, Feng Z, Gilliland G, Bhat TN, Weissig H et al. The protein data bank. 2000. [cited: 2017 Aug 1]. http://www.rcsb.org/pdb/home/home.do
2. Dassault Systemes BIOVIA. BIOVIA Discovery Studio Visualizer, v16.1.0.15350, San Diego: Dassault Systemes, 2015 [cited: 2017 Aug 1]. Available from: http://accelrys.com/products/collaborative-science/biovia-discovery-studio/visualization-download.php

Hot Spots at the Protein-Protein Interface

7

7.1 Interface Hot Spots

Hot spots are favored residues at the protein-protein interfaces for effective binding with optimal energy. These residues contribute to the stability of the protein subunit-subunit complexes. They efficiently distinguish specific from non-specific protein-protein interactions. Hot spots have $\Delta\Delta G$ values in the range of −0.9 to 10 kcal. mol^{-1}. Thus, hot spots ($\Delta\Delta G \geq 1.5$ kcal.mol^{-1}), warm residues (0.5–1.5 kcal.mol^{-1}), and unimportant residues (<0.5 kcal.mol^{-1}) are defined based on energy as described elsewhere [1].

7.2 Alanine-Scanning Energetics Database (ASEdb)

The ASEdb contains energy data for side chain interactions determined by alanine-scanning mutagenesis [2]. It stores information on single alanine mutations in protein-protein complex with experimentally known binding affinities. This is relevant for studying the contributions of single amino acid to protein interactions. Accessible surface areas (ASA) of the mutated side chain in known structures are also computed and stored.

7.3 Interatomic Interface Interaction

Interatomic interaction is defined such that two atoms from two subunits were considered to interact if the distance between their centers is less than the sum of their van der Waals (vdW) radii plus a cutoff distance of 0.5 Å, at which cutoff the mean of side chain-side chain interaction (S_1S_2I) is maximum and their corresponding standard deviation is minimum (Fig. 7.1). It is observed that side chain-side chain interaction (S_1S_2I) dominates at the interface and is used for defining hot spots.

P. Kangueane, C. Nilofer, *Protein-Protein and Domain-Domain Interactions*,
https://doi.org/10.1007/978-981-10-7347-2_7

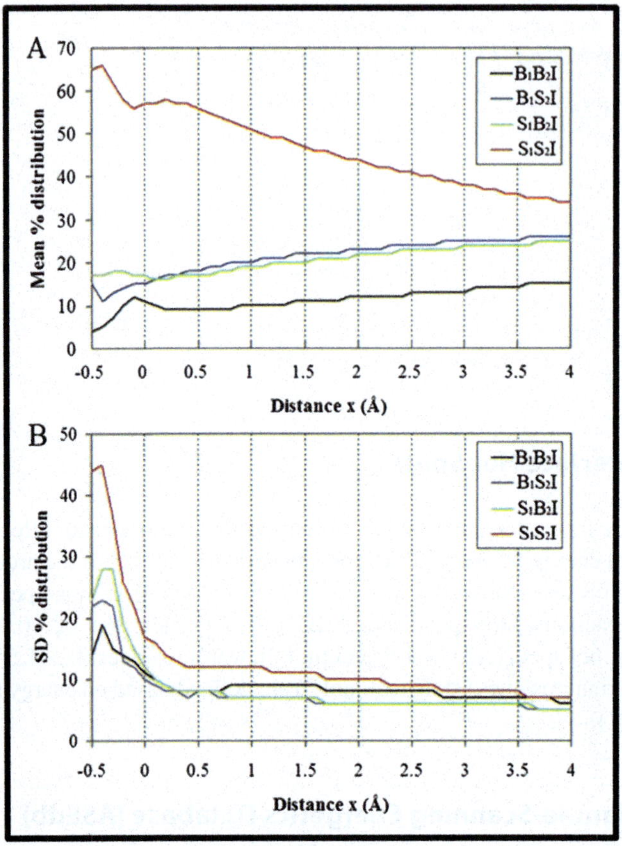

Fig. 7.1 The mean distribution of side chain-side chain interaction (S_1S_2I) dominating at the interface in a selected set of protein-protein complexes is shown. Interatomic interaction is defined such that two atoms from two subunits were considered to interact if the distance between their centers is less than the sum of their van der Waals (vdW) radii plus a cutoff distance of x Å. Mean percentage of side chain-side chain interaction (S_1S_2I) is maximum, and their corresponding standard deviation (SD) is minimum at $x = 0.5$ Å. This image is adapted from Li et al. [4] under the open access creative commons attribution license

7.4 Favorable and Unfavorable Contacts

Protein-protein interaction leads to stable interface for specific biochemical or regulatory function. The interface is made of interatomic bonds with favorable contacts between two subunits as shown in Table 7.1 based on properties described in Table 7.2. An example of favorable contacts is shown in Fig. 7.2.

Table 7.1 Definition of interatomic favorable and unfavorable contacts based on description in Table 7.2

Subunit A										
			I	II	III	IV	V	VI	VII	VIII
	Atom class		Hydrophilic	Acceptor	Donor	Hydrophobic	Aromatic	Neutral	Neutral donor	Neutral acceptor
Subunit B	I	Hydrophilic	+	+	+	–	+	+	+	+
	II	Acceptor	+	–	+	–	+	+	+	–
	III	Donor	+	+	–	–	+	+	–	+
	IV	Hydrophobic	–	–	–	+	+	+	+	+
	V	Aromatic	+	+	+	+	+	+	+	+
	VI	Neutral	+	+	+	+	+	+	+	+
	VII	Neutral donor	+	+	–	+	+	+	–	+
	VIII	Neutral acceptor	+	–	+	+	+	+	+	–

+ = favorable contacts; – = unfavorable contacts

Table 7.2 Description of amino acid atom level chemical properties as described elsewhere [3]

Class	Property name	Description
I	Hydrophilic	N or O atoms that can donate and accept H bonds
II	Acceptor	N or O atoms that can only accept a H bond
III	Donor	N that can only donate a H bond
IV	Hydrophobic	C atoms that are not in aromatic rings and do not have a covalent bond to a hydrophilic atom
V	Aromatic	C atoms in aromatic rings
VI	Neutral	C atoms that have a covalent bond to at least one atom of class I or two or more atoms from class II or III; N atoms if it has covalent bonds with 3 C; S atoms in all cases
VII	Neutral donor	C atoms that has a covalent bond with only one atom of class III
VIII	Neutral acceptor	C atoms that has covalent bond with only one atom of class II

C carbon, *N* nitrogen, *S* sulfur, *O* oxygen

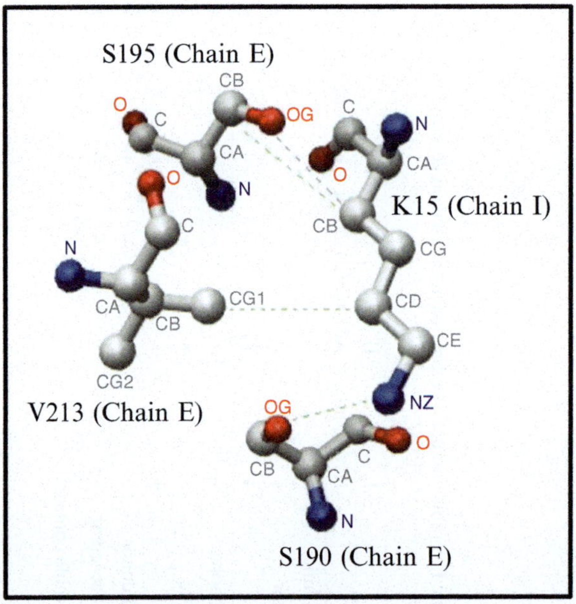

Fig. 7.2 Illustration of favorable and unfavorable contacts is shown. The interaction of residue K15 (PDB ID: BPTI, Chain I) to residues S190, S195, and V213 (trypsin, Chain E) is shown (PDB ID: 2PTC). K15 has three interacting side chain atoms (CB, CD, and NZ). It should be noted that these three atoms are involved in favorable contacts (green line) and only CB participates in unfavorable contacts (red line). Carbon atom, white; oxygen atom, red; nitrogen atom, blue (This image is adapted from Li et al. [4] under the open access creative commons attribution license)

Table 7.3 List of databases for interface hot spots

S. No.	Name	Description	URL or statistics	Year	References
1	PCRPi-DB	Presaging critical residues in protein interfaces database	http://www.bioinsilico.org/PCRPIDB	2011	[5]
2	HotSprint	Predicted	http://prism.ccbb.ku.edu.tr/hotsprint	2008	[6]
3	BID	Binding interface database	http://tsailab.org/BID/	2003	[7]

Table 7.4 Hot spot prediction methods

S. no	Server	Prediction methods	Year	References
1	APIS	Combination of protrusion index with solvent accessibility	2010	[8]
2	–	Multiple criteria linear programming	2011	[9]
3	DBAC	Deeply buried atomic contacts	2011	[10]
4	–	Random forest model with hybrid features	2012	[11]
5	–	Structural neighborhood properties	2013	[12]
6	–	Machine learning using spatial neighbor residues	2014	[13]
7	–	Ensemble learning method	2017	[14]

7.5 Databases for Interface Hot Spots

Protein-protein interaction (PPI) is important for the understanding of cell and molecular biology events. Hot spots are important for describing PPI. Collection, gleaning, and archive of interface hot spots in the form of databases (Table 7.3) are useful functional genomics through genome annotations.

7.6 Interface Hot Spot Prediction

Protein-protein interaction is an important phenomenon in complex cellular function. The protein-protein binding is described using hot spots characteristics of favorable contacts. Therefore, prediction of hot spots at the protein-protein interface is of significance. We describe currently available methods to predict hot spots (Table 7.4). Hot spot prediction is now possible using features such as protrusion index with solvent accessibility [8], interface criteria [9], buried atomic contacts [10], hybrid features [11] and structural properties [12] using linear programming [12], random forest model [11], and machine learning models [13–14].

Exercises

1. Define protein interface hot spots.
2. Illustrate favorable contacts using an example.

3. Illustrate side chain-side chain interaction mean distribution at the interface.
4. Name some hot spots database.
5. Name some hot spot prediction models.
6. Discuss improvements in hot spot prediction.
7. Expand ASEdb.
8. What are the energy range for hot spots, warm residues, and unimportant residues?
9. State the atom level chemical properties used to describe favorable contacts.
10. Illustrate interatomic favorable and unfavorable contacts in matrix format.

References

1. Gao Y, Wang R, Lai L. Structure-based method for analyzing protein-protein interfaces. J Mol Model. 2004;10(1):44–54.
2. Thorn KS, Bogan AA. ASEdb: a database of alanine mutations and their effects on the free energy of binding in protein interactions. Bioinformatics. 2001;17(3):284–5.
3. Li L, Zhao B, Cui Z, Gan J, Sakharkar MK, Kangueane P. Identification of hot spot residues at protein-protein interface. Bioinformation. 2006;1(4):121–6.
4. Sobolev V, Wade RC, Vriend G, Edelman M. Molecular docking using surface complementarity. Proteins. 1996;25(1):120–9.
5. Segura J, Fernandez-Fuentes N. PCRPi-DB: a database of computationally annotated hot spots in protein interfaces. Nucleic Acids Res. 2011;39(Database issue):D755–60.
6. Guney E, Tuncbag N, Keskin O, Gursoy A. HotSprint: database of computational hot spots in protein interfaces. Nucleic Acids Res. 2008;36(Database issue):D662–6.
7. Fischer TB, Arunachalam KV, Bailey D, Mangual V, Bakhru S, Russo R, et al. The binding interface database (BID): a compilation of amino acid hot spots in protein interfaces. Bioinformatics. 2003;19(11):1453–4.
8. Xia JF, Zhao XM, Song J, Huang DSAPIS. Accurate prediction of hot spots in protein interfaces by combining protrusion index with solvent accessibility. BMC Bioinform. 2010;11:174.
9. Chen R, Zhang Z, Wu D, Zhang P, Zhang X, Wang Y, et al. Prediction of protein interaction hot spots using rough set-based multiple criteria linear programming. J Theor Biol. 2011;269(1):174–80.
10. Li Z, Wong L, Li J. DBAC: a simple prediction method for protein binding hot spots based on burial levels and deeply buried atomic contacts. BMC Syst Biol. 2011;5(Suppl 1): S5.
11. Wang L, Liu ZP, Zhang XS, Chen L. Prediction of hot spots in protein interfaces using a random forest model with hybrid features. Protein Eng Des Sel. 2012;25(3):119–26.
12. Deng L, Guan J, Wei X, Yi Y, Zhang QC, Zhou S. Boosting prediction performance of protein-protein interaction hot spots by using structural neighborhood properties. J Comput Biol. 2013;20(11):878–91.
13. Wang L, Zhang W, Gao Q, Xiong C. Prediction of hot spots in protein interfaces using extreme learning machines with the information of spatial neighbor residues. IET Syst Biol. 2014;8(4):184–90.
14. SS H, Chen P, Wang B, Li J. Protein binding hot spots prediction from sequence only by a new ensemble learning method. Amino Acids. 2017;

Principles of Protein-Protein Interaction

8

8.1 Protein Size, Shape, and Complementation

Proteins have irregular size and shapes as shown in Fig. 8.1. Protein-protein interfaces are subsequently irregular in nature. Interfaces are formed between different types of protein partners. They have different modes of protein-protein binding as shown in Fig. 8.1. Protein-protein complementation in protein-protein binding is illustrated in Fig. 8.1.

8.2 Protein-Protein Interfaces

Protein-protein interfaces (Fig. 8.2) are critical for molecular function such as enzyme catalysis, regulation, immune function, and inhibition. Enzyme catalysis and regulation are obligatory (essential) in nature. Inhibition is largely non-obligatory (not essential) in nature. It should be noted that some are characterized as feedback inhibition with obligatory features.

8.3 Interface, Surface, and Core in Complexes

A protein-protein complex is characterized by three regions (core, surface, and interface) as shown in Figs. 8.3 and 8.4. The surface region is accessible to water, while the core region is inaccessible to water. Thus, the accessible surface area (ASA) will be zero for residues in the core region. The interface is that part of the region in the surface before complex formation. This region becomes inaccessible to water upon complex formation. The difference in property between interface and surface in a complex state is the driving force for protein-protein binding. It should be noted that the interface is characterized by irregular geometry with specific chemical features required for binding.

P. Kangueane, C. Nilofer, *Protein-Protein and Domain-Domain Interactions*,
https://doi.org/10.1007/978-981-10-7347-2_8

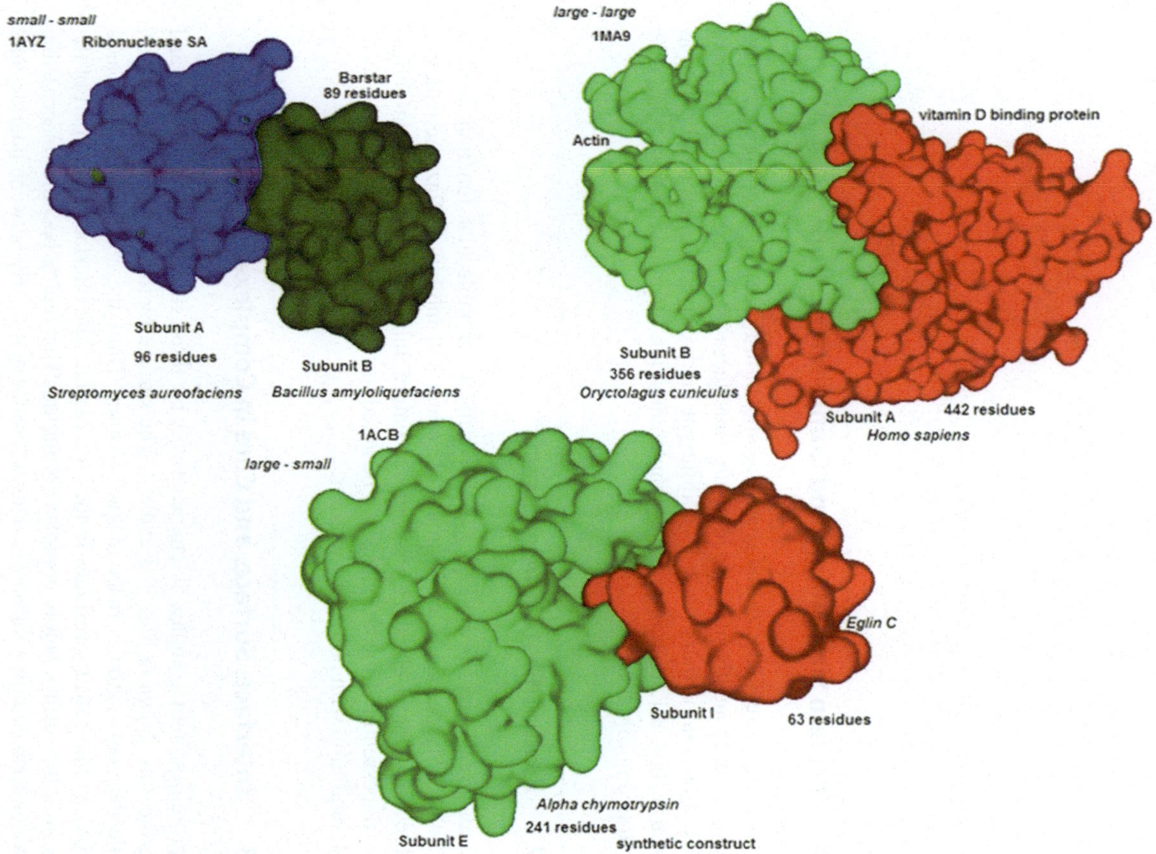

Fig. 8.1 Protein size, shape, and complementation in protein-protein complexes. Proteins in complexes have dissimilar sizes and irregular shapes and complement with each other (This figure is adapted from Vaishnavi et al. [1] under the open access creative commons attribution license)

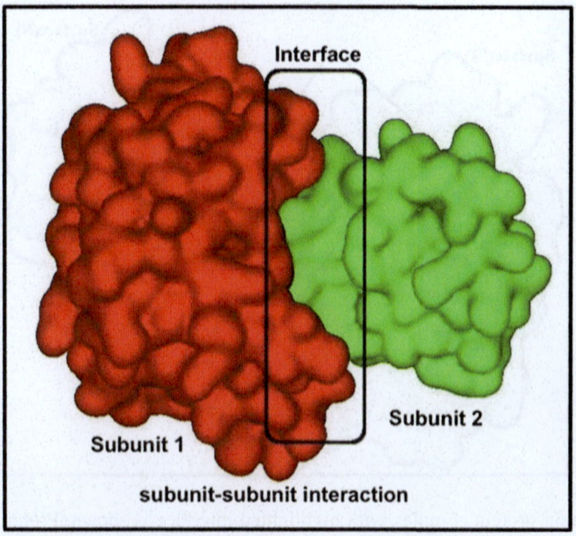

Fig. 8.2 A protein-protein complex between two subunits (subunit 1 and subunit 2) showing a protein-protein interface. The interface is irregular and complementary in nature

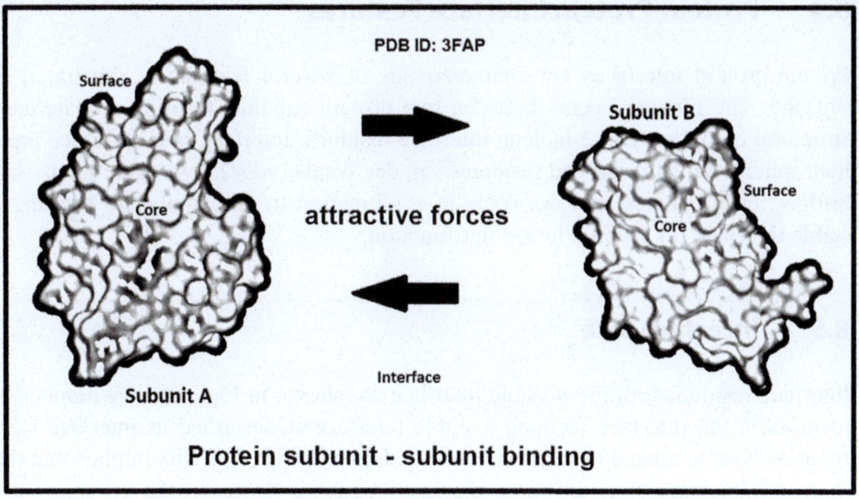

Fig. 8.3 An illustration of protein subunit-subunit binding is shown. Proteins have irregular surface. The three regions (core, surface, and interface) are shown. Core is buried and inaccessible to water. Interface is that part of the surface in each of the interacting partners that is involved in subunit-subunit protein binding. The difference in physicochemical properties of this region of the surface with the rest of the surface is realized

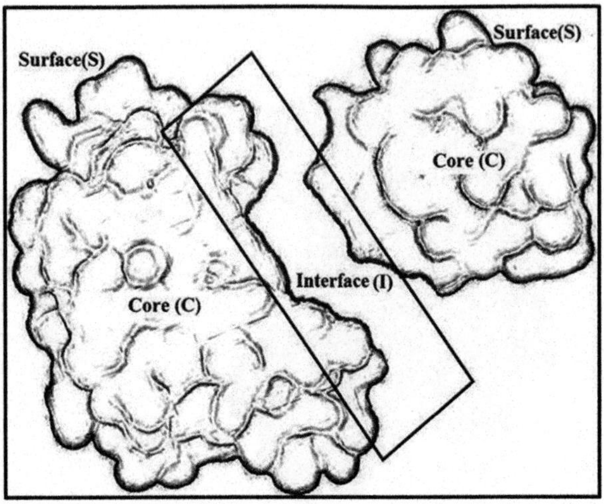

Fig. 8.4 A protein-protein complex with highlighted interface is shown. Protein-protein binding leads to a stable interface to help in catalysis, regulation, and immune response (This figure is adapted from Sowmya et al. [2] under the open access creative commons attribution license)

8.4 Protein-Protein Interface Features

Protein-protein interfaces are characteristics of several features as illustrated in Fig. 8.5. The binding occurs between two protein subunits forming a quaternary structural complex. These include interface residues, interface size, interface area, hydrophobic effects, charged residues, van der Waals (vdW), hydrogen bonds, salt bridges, electrostatics, and hot spots in combination to one another in forming a stable interface for specific molecular function.

8.5 Interface Size

Interface residues forming a stable interface are shown in Fig. 8.6. The number of residues at the interface forming a stable interface is described as interface size. Interface size is related to interface area as shown in Fig. 8.7. This implies that the more the interface size, the more the interface area. However, the interfaces are either small or medium or large in size as illustrated with examples in Fig. 8.8.

8.6 Interface Hydrophobic Effects

Analysis of protein-protein complexes [8] shows hydrophobic interface with closely packed hydrophobic residues (Fig. 8.9). However, it was further asserted that interfaces are more hydrophobic than surface [9]. Nonetheless, interfaces are less

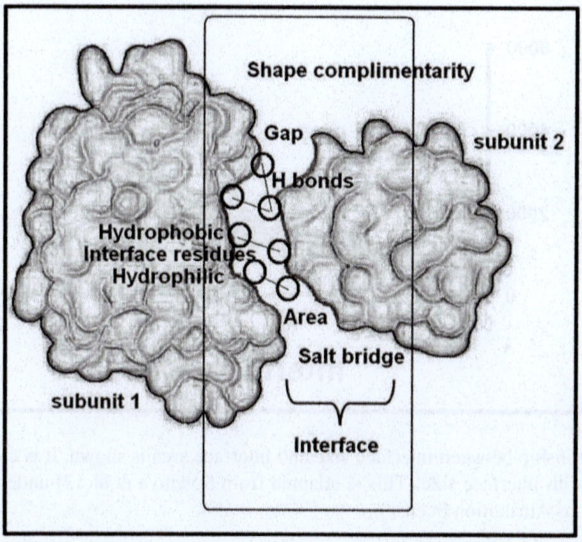

Fig. 8.5 An illustration of a protein-protein interface with all possible interface features. Interface features include interface residues, shape complementation, gap volume, interface area, interface size, hydrogen bonds, salt bridges, hydrophobic residues, and hydrophilic residues

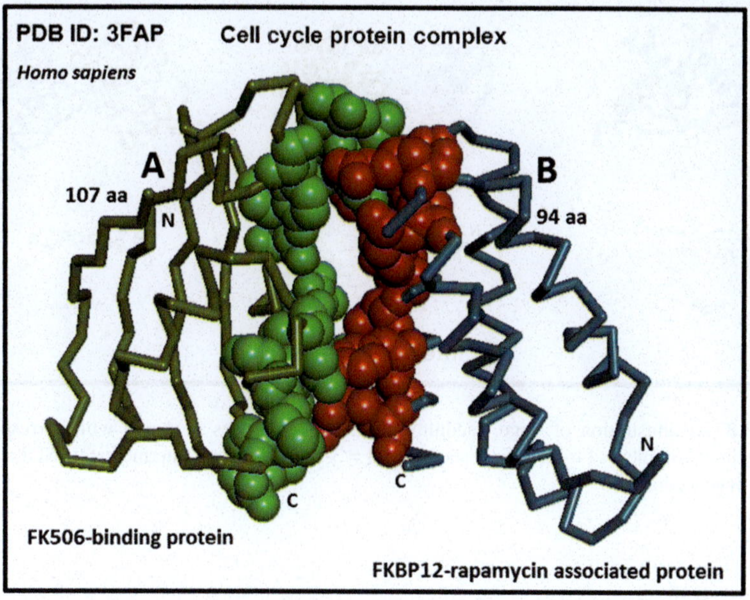

Fig. 8.6 An illustration of interface residues at the protein-protein interface using CPK representation is illustrated. This shows that a stable interface is critical for protein-protein binding (This figure is adapted from Nilofer et al. [1] under the open access creative commons attribution license)

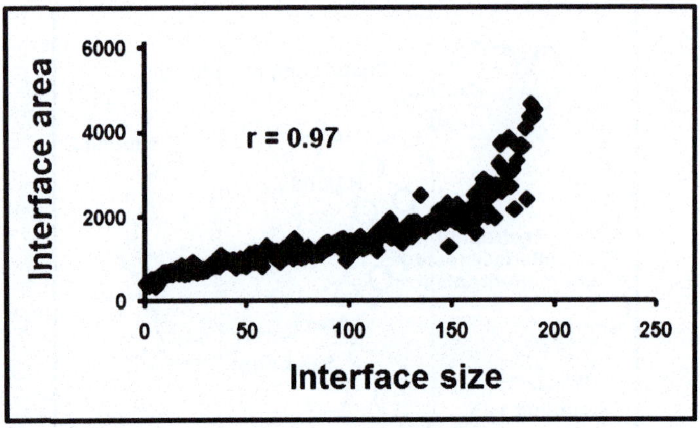

Fig. 8.7 Relationship between interface size and interface area is shown. It is clear that interface area increases with interface size (This is adapted from Sowmya et al. [2] under the open access creative commons Attribution license)

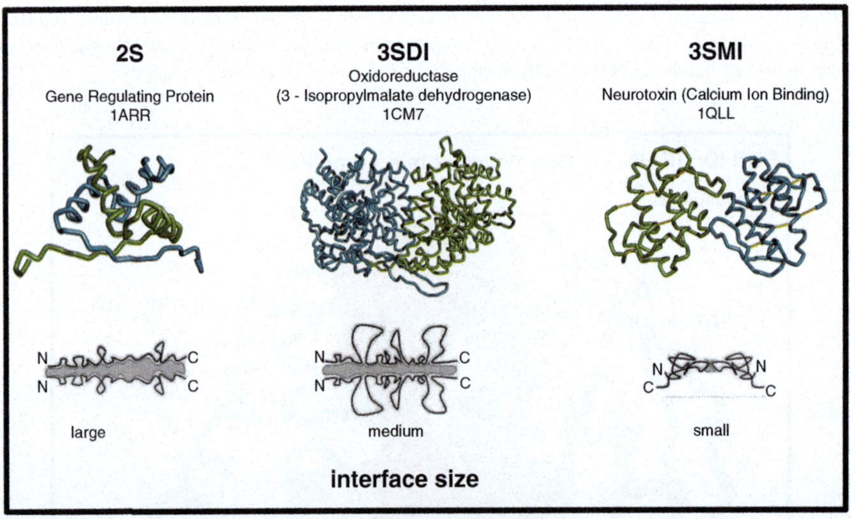

Fig. 8.8 An illustration of large, medium, and small interfaces is shown with corresponding homodimer complexes [4–7]. 2S = 2-state; 3SMI = 3-state monomer intermediate; 3SDI = 3-state dimer intermediate

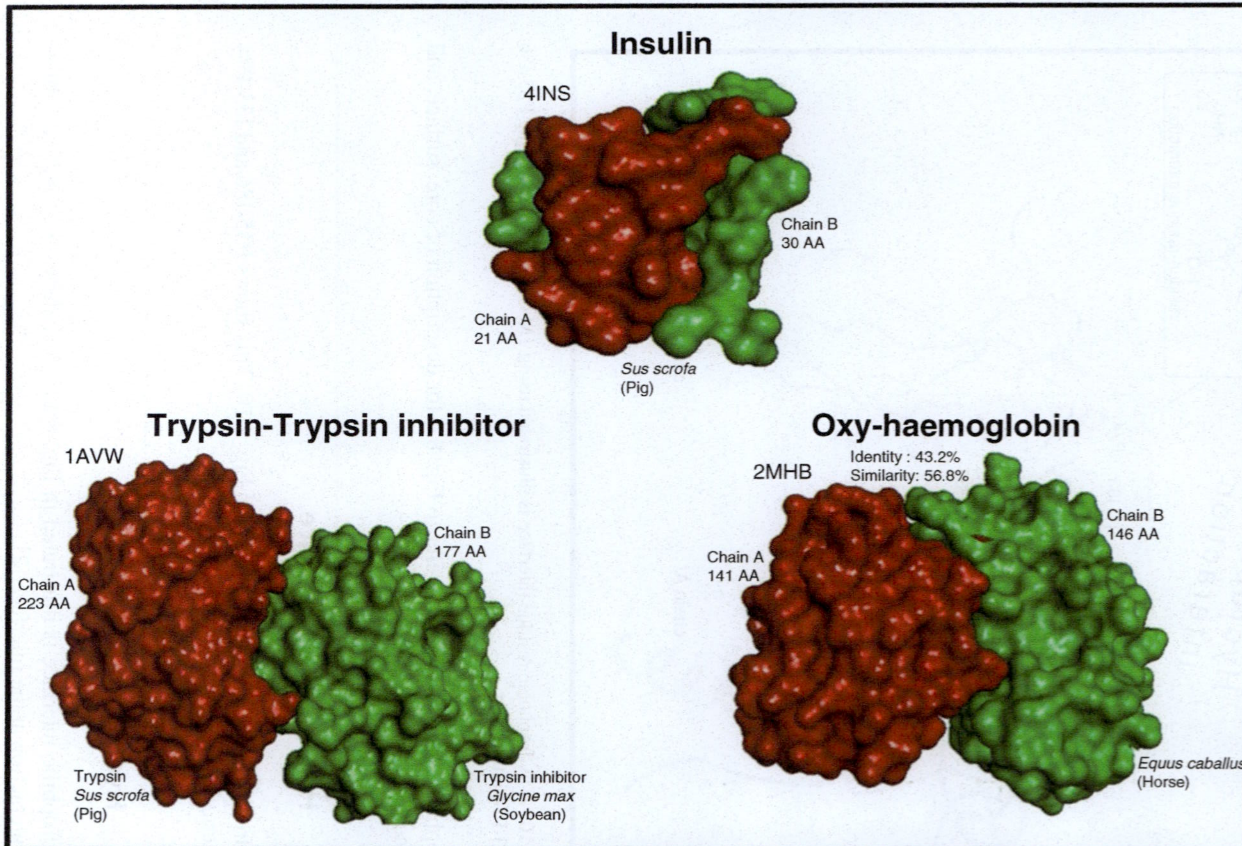

Fig. 8.9 A dataset of three protein-protein complexes (insulin, trypsin-trypsin inhibitor, and oxyhemoglobin) having hydrophobic interface [8] is shown. It should be noted that subunits in oxyhemoglobin are homologous with 57% sequence similarity and hence it is a homodimer. The subunits in insulin (<10% sequence similarity) and trypsin-trypsin inhibitor (negligible sequence similarity) complexes are non-homologous, and hence they are heterodimers

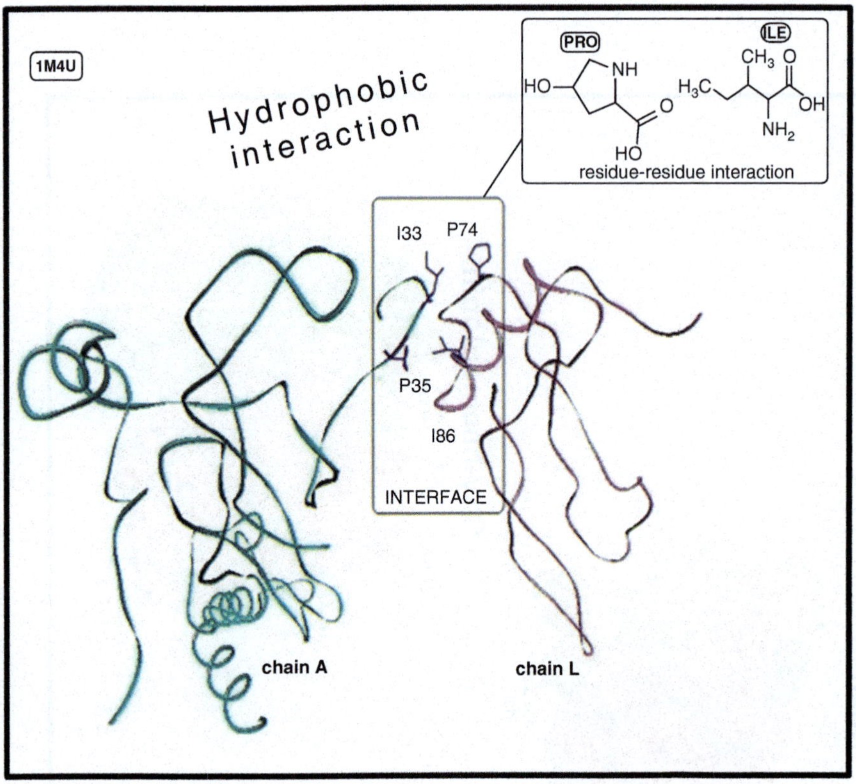

Fig. 8.10 A hydrophobic residue interface is illustrated using 1M4U (PDB ID) showing P35-I86 and I33-P74 interaction

hydrophobic than core. Please refer to Fig. 8.4 for description of core, surface, and interface.

8.7 Hydrophobic Interface

A hydrophobic interface is illustrated in Fig. 8.10. This shows P35-I86 and I33-P74 interaction at the interface [2].

8.8 Hydrophilic Interface

A hydrophilic interface is illustrated in Fig. 8.11. This shows N61-N20 and R63-R15 interaction at the interface [2].

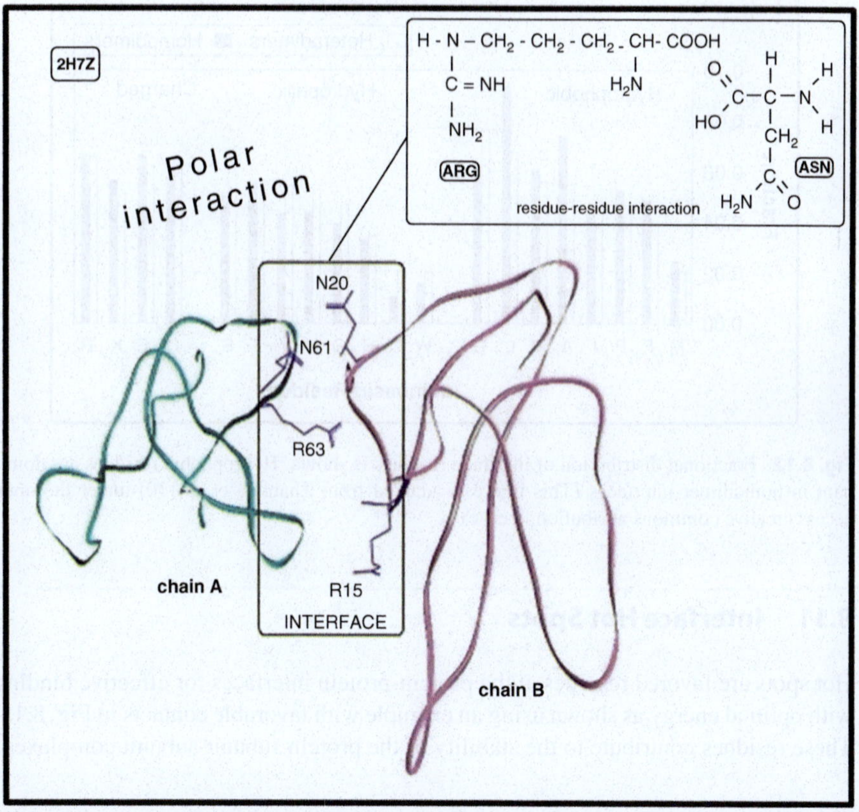

Fig. 8.11 A hydrophilic residue interface is illustrated using 2H7Z (PDB ID) showing N61-N20 and R63-R15 interaction

8.9 Interface Residues in Homo and Hetero dimer

Residue level analysis of interface shows the presence of hydrophobic, hydrophilic, and charged residues in both homodimer and heterodimer complexes as shown in Figs. 8.12 and 8.13. Hydrophobic residues are dominant in homodimer interfaces unlike in heterodimer interfaces (Fig. 8.12). The ratio of charged residues at the interface to interior is high for heterodimer protein complexes (Fig. 8.13). This trend is true for hydrophobic residues in homodimer protein complexes.

8.10 Interface Side Chain-Side Chain Interaction

Protein-protein are dominant with side chain-side chain interatomic interaction at the interface as shown in Fig. 8.14. It should be noted that backbone-backbone and side chain-backbone interatomic interactions are subdominant at the interface.

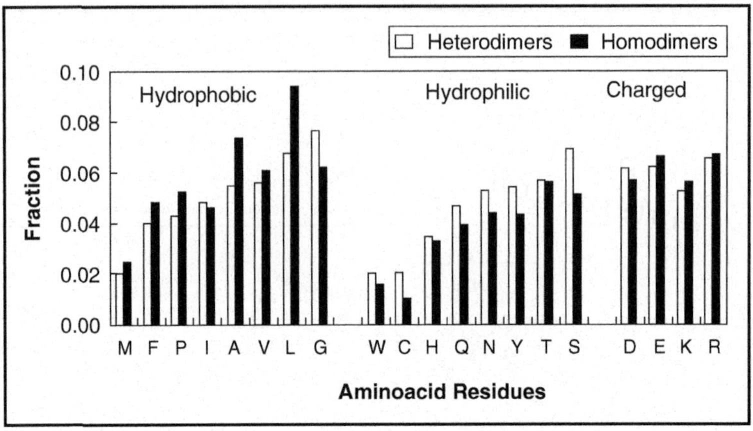

Fig. 8.12 Fractional distribution of interface residues is shown. Hydrophobic residues are dominant in homodimer interfaces (This figure is adapted from Zhanhua et al. [10] under the open access creative commons attribution license)

8.11 Interface Hot Spots

Hot spots are favored residues at the protein-protein interfaces for effective binding with optimal energy as shown using an example with favorable contacts in Fig. 8.15. These residues contribute to the stability of the protein subunit-subunit complexes.

8.12 Interface Hydrogen Bonds and Salt Bridges

The presence of positive (blue), negative (red), and neutral (green) residues at the protein-protein interface is shown in Fig. 8.16. The presence of aliphatic and aromatic residues is also highlighted. The interface is also made of non-bonded contacts, hydrogen bonds, and salt bridges. Moreover, hydrogen bonds at the interface increase with interface size in both homo- and heterodimer complexes as shown in Fig. 8.17.

8.13 Interface Binding Energy

The interface total binding energy increases with interface size in obligatory, non-obligatory, and immune complexes as shown in Fig. 8.18. Thus, binding energy increases with interface size. More the interface size, high is the binding energy. This total energy is made of vdW energy, hydrogen bond energy, and electrostatic energy.

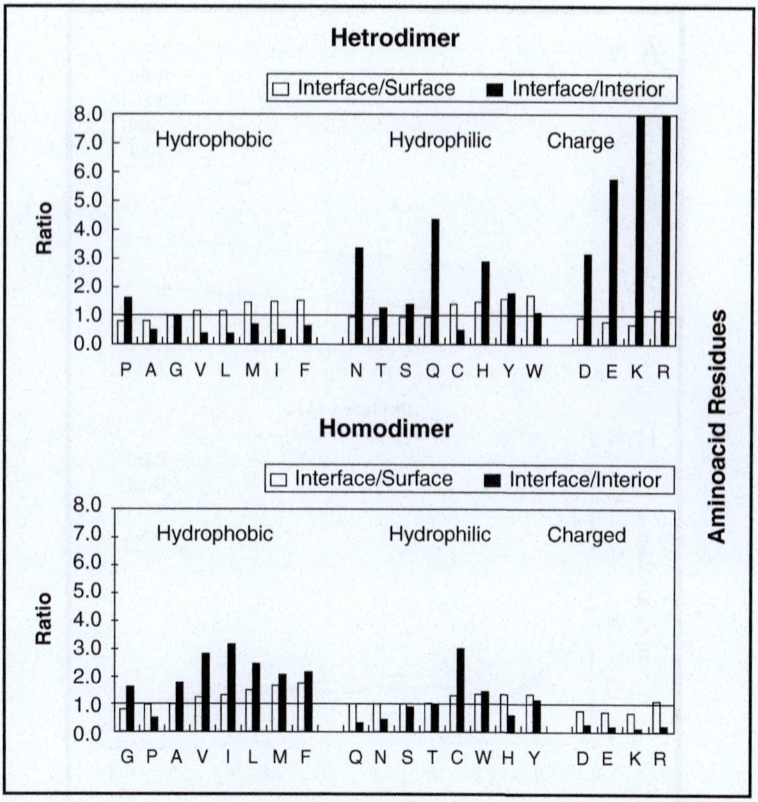

Fig. 8.13 The distribution of amino acid residues as a ratio of interface to surface and interior is shown for heterodimer and homodimer protein complexes. The ratio of charged residues at the interface to interior is high for heterodimer protein complexes. This trend is true for hydrophobic residues in homodimer protein complexes (This figure is adapted from Zhanhua et al. [10] under the open access creative commons attribution license)

8.14 Interface Van Der Waals (vdW)

The contribution by van der Waals (vdW) energy at the interface of obligatory, non-obligatory, and immune complexes is about 75% ± 11% as shown in Fig. 8.19. Moreover, vdW increases with interface size in obligatory, non-obligatory, and immune complexes as shown in Fig. 8.20. It should be noted that this relationship is less in immune complexes compared to obligatory and non-obligatory complexes.

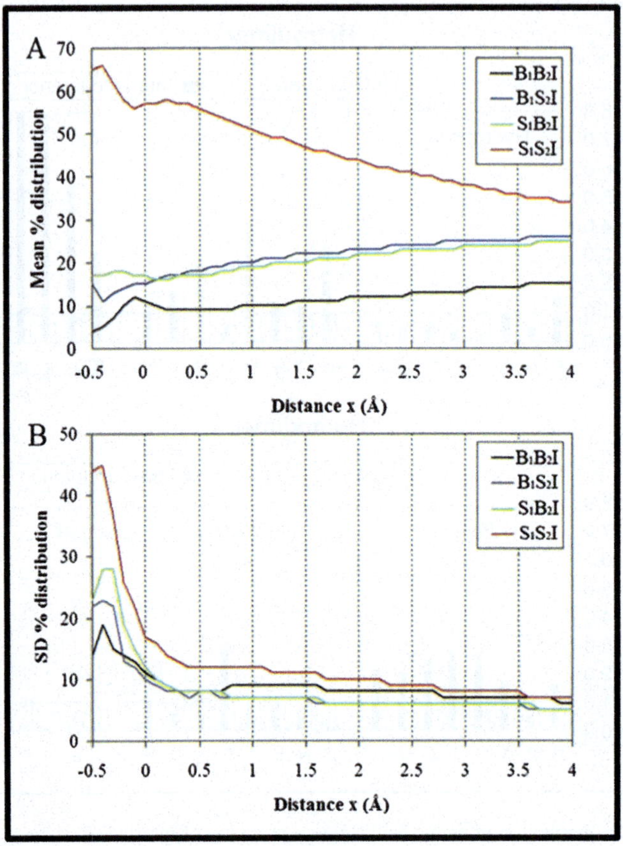

Fig. 8.14 Distribution of side chain-side chain interaction (S_1S_1I) at the interface is shown as a function of distance x (Å). Two atoms are considered to be interacting of the interacting if the distance between them is within the sum of their vdW radii plus x distance (This image is adapted from Li et al. [11] under the open access creative commons Attribution license)

8.15 Interface Hydrogen Bonds

The contribution by hydrogen bond energy at the interface of obligatory, non-obligatory, and immune complexes is about 15% ± 6.5% as shown in Fig. 8.21. It is observed that hydrogen bond contribution in immune interfaces is larger than obligatory and non-obligatory interfaces. Moreover, hydrogen bonds reasonably increase with interface size in obligatory and immune interfaces as shown in Fig. 8.22. However, this is not true with non-obligatory interfaces.

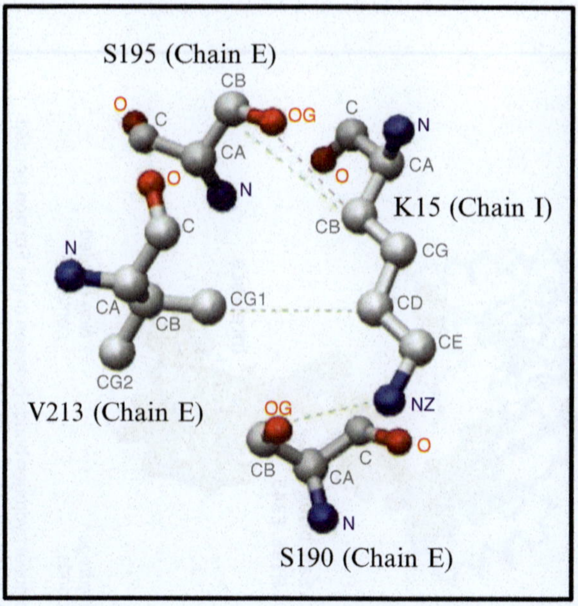

Fig. 8.15 An interface hot spot is shown. The interaction of residue K15 (PDB ID: BPTI, Chain I) to residues S190, S195, and V213 (trypsin, Chain E) is shown (PDB ID: 2PTC). K15 has three interacting side-chain atoms (CB, CD and NZ). It should be noted that these three atoms are involved in favorable contacts (green line) and only CB participates in unfavorable contacts (red line). Carbon atom, white; oxygen atom, red; nitrogen atom, blue (This image is adapted from Li et al. [11] under the open access creative commons attribution license)

8.16 Interface Electrostatics

The contribution by electrostatic energy at the interface of obligatory, non-obligatory, and immune complexes is about 11.3% ± 8.7% as shown in Fig. 8.23. Moreover, electrostatic energy fairly increases with interface size among non-obligatory complexes as shown in Fig. 8.24.

8.17 Gene Fusion and Fusion Protein

Gene fusion is an important phenomenon in molecular evolution resulting in the formation of fusion proteins as shown in Fig. 8.25. Thus, an example of subunit-subunit interaction in *Thermus thermophilus* evolved into a chimeric protein with a domain-domain interface in *Saccharomyces cerevisiae* is shown using IGPS. A linker connects the two domains for conserved molecular function.

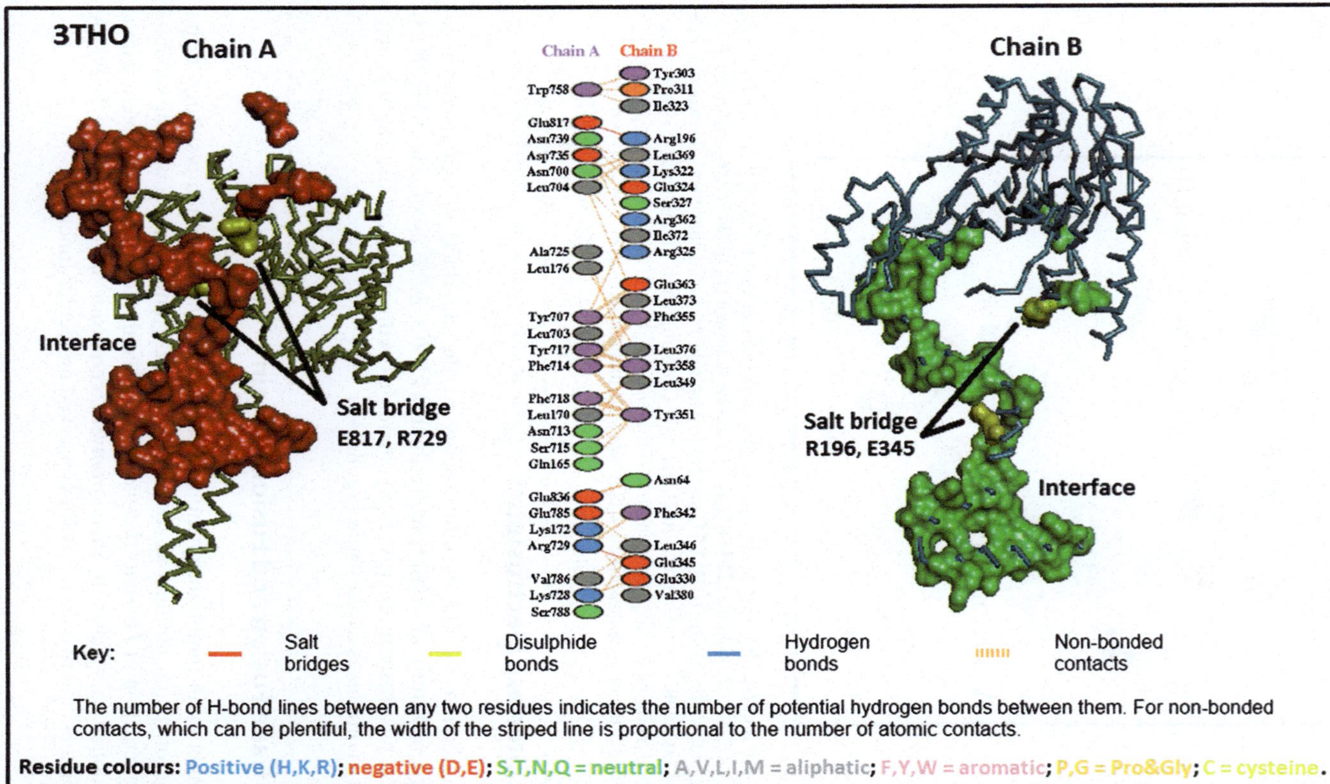

Fig. 8.16 Interface interaction between two chains is shown using PDBsum (http://www.ebi.ac.uk/thornton-srv/databases/cgi-bin/pdbsum/GetPage.pl?pdbcode = index.html) with an example of regulatory protein (PDB ID: 3THO). Interface residues of chain A and chain B are colored red and green, respectively. Interface salt bridge interaction is indicated in red color between chains with labeled residues

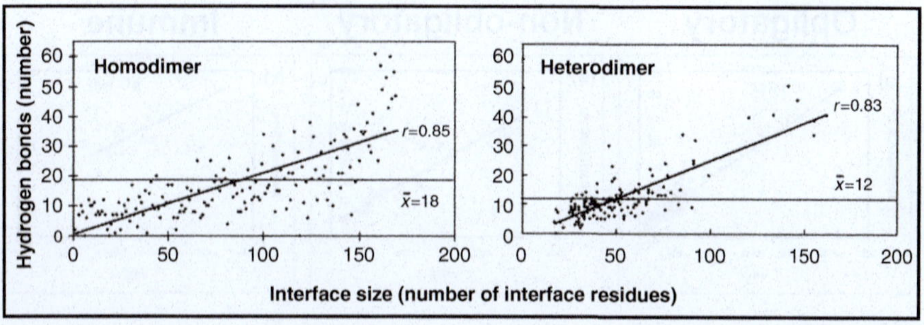

Fig. 8.17 Relationship between interface size and proportion of hydrogen bonds is shown for homodimer and heterodimer interfaces (This is adapted from Zhanhua et al. [10] under the open access creative commons attribution license)

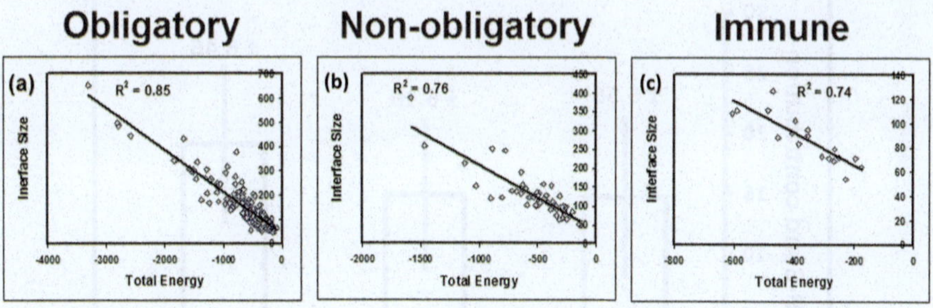

Fig. 8.18 Relationship between interface size and energy in obligatory, non-obligatory, and immune complexes is shown. Binding energy increases with interface size (This image is adapted from Nilofer et al. [3] under the open access creative commons attribution license)

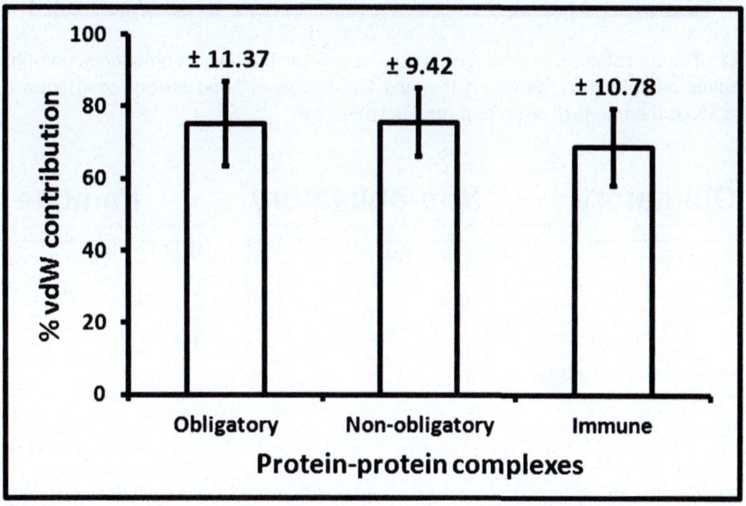

Fig. 8.19 The contribution by van der Waals (vdW) energy at the interface of obligatory, non-obligatory, and immune complexes is shown. It is noted that vdW contributes to about 75% ± 11% at the interface of protein-protein complexes

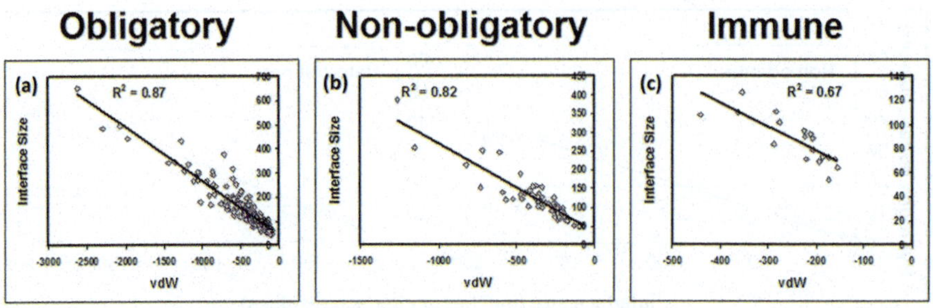

Fig. 8.20 Relationship between interface size and vdW energy at the interface of obligatory, non-obligatory, and immune complexes is shown (This image is adapted from Nilofer et al. [3] under the open access creative commons attribution license)

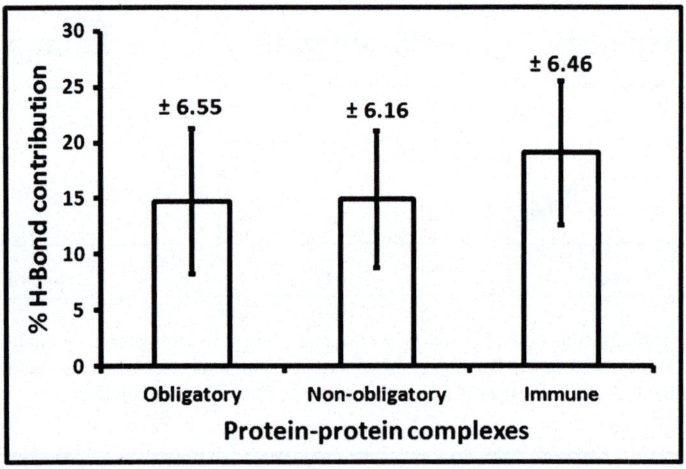

Fig. 8.21 The contribution by hydrogen bond energy at the interface of obligatory, non-obligatory, and immune complexes is shown. It is noted that hydrogen bond energy contributes to about 15% ± 6.5% at the interface of protein-protein complexes

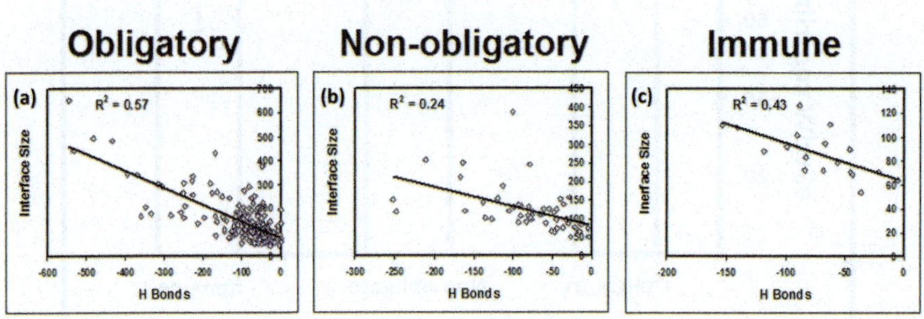

Fig. 8.22 Relationship between interface size and hydrogen bond energy at the interface of obligatory, non-obligatory, and immune complexes is shown (This image is adapted from Nilofer et al. [3] under the open access creative commons attribution license)

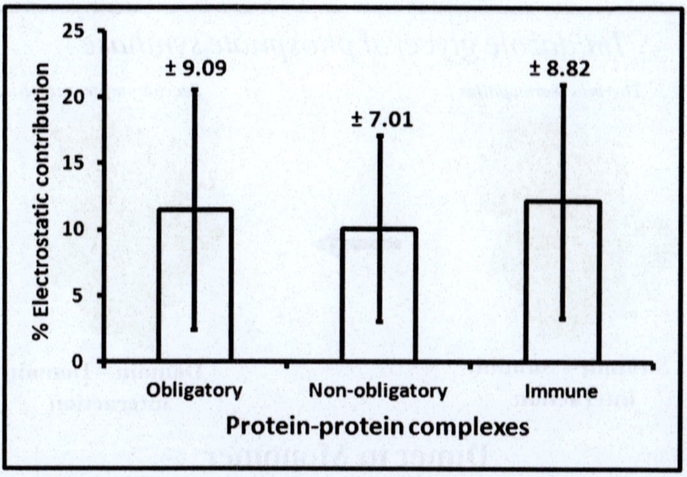

Fig. 8.23 The contribution by electrostatic energy at the interface of obligatory, non-obligatory, and immune complexes is shown. It is noted that electrostatic energy contributes to about 11.3% ± 8.7% at the interface of protein-protein complexes

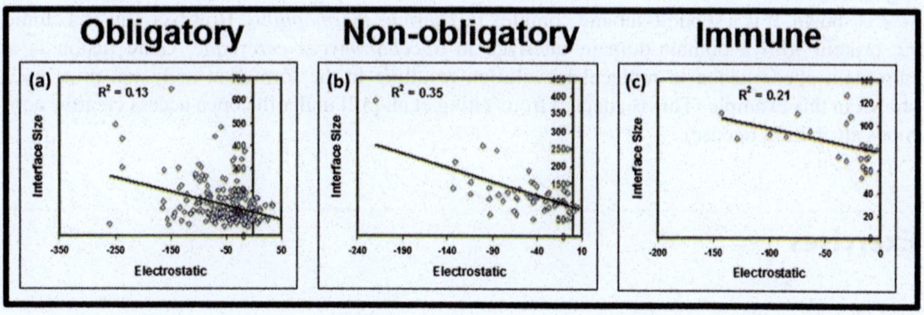

Fig. 8.24 Relationship between interface size and electrostatic energy at the interface of obligatory, non-obligatory, and immune complexes is shown (This image is adapted from Nilofer et al. [3] under the open access creative commons attribution license)

8.18 Conclusion

The driving force for protein-protein interaction in the formation of a stable interface is described using interface features in this chapter. Protein-protein binding is deterministic of interface size constituted by interface residues. The strength of binding is proportional to interface area contributed by interface size. It is known that the interface is van der Waals (vdW) dominant with selective hydrogen bonds and (or) electrostatics. These features work in combination during protein-protein interaction in the formation of a stable interface for specific molecular function. It should be noted that the role played by fusion proteins while conserving molecular function through gene fusion is contextual.

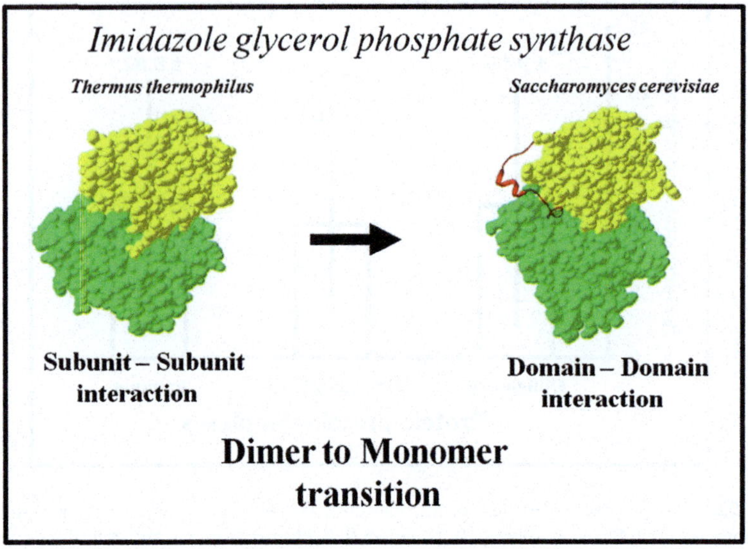

Fig. 8.25 A structural illustration of IGPS in *Thermus thermophilus* and *Saccharomyces cerevisiae* is shown. It is a subunit-subunit complex in *Thermus thermophilus.* However, this is a chimeric protein with a domain-domain interface in *Saccharomyces cerevisiae.* Gene fusion is an important phenomenon in molecular evolution resulting in the formation of fusion proteins as shown in this example (This is adapted from Yiting et al. [12] under the open access creative commons attribution license)

Exercises

1. Illustrate protein size, shape, and complementation in complexes.
2. Illustrate protein-protein interface.
3. Illustrate protein-protein interface features.
4. State any ten interface features.
5. Expand vdW.
6. What is the contribution made by vdW at the interface?
7. Explain interface hydrophobic effect.
8. Illustrate hydrogen bonds at the interface.
9. Illustrate electrostatics at the interface.
10. Show hydrogen bonds and salt bridges at the interface.
11. Show the relationship between interface size and area.
12. What is interface size?
13. What is interface area?
14. Show the relationship between total energy and interface size.
15. Show the relationship between vdW energy and interface size.
16. Give the dominant features at the interface.
17. Show domain-domain interface.

18. What is gene fusion?
19. What are obligatory complexes?
20. What are non-obligatory complexes?

References

1. Vaishnavi A, Sowmya G, Kalaivanii J, Ilakya S, Kangueane U, Kangueane P. Interaction modes at protein hetero-dimer interfaces. Bioinformation. 2010;4(7):310–9.
2. Sowmya G, Anita S, Kangueane P. Insights from the structural analysis of protein heterodimer interfaces. Bioinformation. 2011;6(4):137–43.
3. Nilofer C, Sukhwal A, Mohanapriya A, Kangueane P. Protein-protein interfaces are vdW dominant with selective H-bonds and (or) electrostatics towards broad functional specificity. Bioinformation. 2017;13(6):164–73.
4. Lulu S, Suresh A, Karthikraja V, Arumugam M, Kayathri R, Kangueane P. Structural features for homodimer folding mechanism. J Mol Graph Model. 2009;28(2):88–94.
5. Suresh A, Karthikraja V, Lulu S, Kangueane U, Kangueane P. A decision tree model for the prediction of homodimer folding mechanism. Bioinformation. 2009;4(5):197–205.
6. Karthikraja V, Suresh A, Lulu S, Kangueane U, Kangueane P. Types of interfaces for homodimer folding and binding. Bioinformation. 2009;4(3):101–11.
7. Li L, Gunasekaran K, Gan JG, Zhanhua C, Shapshak P, Sakharkar MK, et al. Structural features differentiate the mechanisms between 2S (2 state) and 3S (3 state) folding homodimers. Bioinformation. 2005;1(2):42–9.
8. Chothia C, Janin J. Principles of protein-protein recognition. Nature. 1975;256(5520):705–8.
9. Jones S, Thornton JM. Protein-protein interactions: a review of protein dimer structures. Prog Biophys Mol Biol. 1995;63:31–65.
10. Zhanhua C, Gan JG, Lei L, Sakharkar MK, Kangueane P. Protein subunit interfaces: heterodimers versus homodimers. Bioinformation. 2005;1(2):28–39.
11. Li L, Zhao B, Cui Z, Gan J, Sakharkar MK, Kangueane P. Identification of hot spot residues at protein-protein interface. Bioinformation. 2006;1(4):121–6.
12. Yiting Y, Lei L, Sakharkar MK, Kangueane P. Insight into gene fusion from molecular dynamics simulation of fused and un-fused IGPS (Imidazole Glycerol Phosphate Synthetase). Bioinformation. 2006;1(3):99–104.

Databases for Protein-Protein Interaction

9

9.1 PPI Databases

Databases play a major role in biomedical drug discovery and development. PPI database stores information related to proteins and their interaction data with other proteins. Data archived include binding affinity, interaction features (physical or chemical), interface, energy, annotations, hot spots, domains, and pathways. There are various categories of PPI databases as shown in Fig. 9.1. Detailed information on general PPI (Table 9.1), PPI for model organism (Table 9.2), PPI networks (Table 9.3), structural data for PPI (Table 9.4), and PPI pathways (Table 9.5) are presented.

9.2 General PPI Database

The understanding of PPI uses different types of data. Therefore, it is important to collect, glean, and archive data useful in the analysis and understanding of PPI. There are 14 PPI databases containing general data on PPI as given in Table 9.1.

9.2.1 DIP

Database of Interacting Proteins [1] discusses PPI's experimental data.

9.2.2 MINT

Molecular INTeraction Database [2] is a public repository on physical interaction among proteins.

P. Kangueane, C. Nilofer, *Protein-Protein and Domain-Domain Interactions*,
https://doi.org/10.1007/978-981-10-7347-2_9

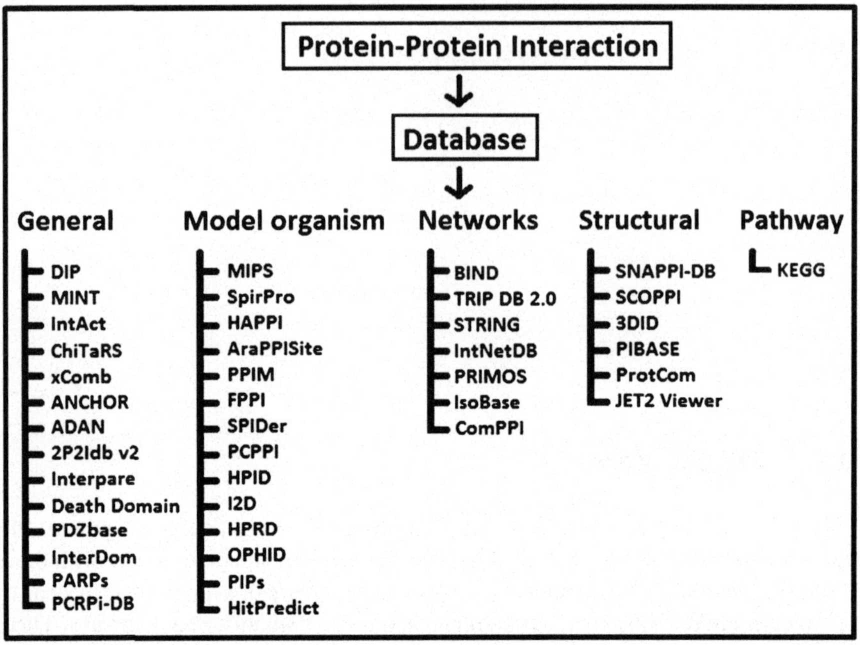

Fig. 9.1 Classification of PPI databases into subcategories is shown. There are a number of databases that archive data describing PPI. The currently available databases are grouped into general, networks, structural, model organism, and pathway. It should be noted that KEGG contain data associated with human diseases such as cancer, viral pathogenesis, and others having information related to PPI

9.2.3 IntAct

IntAct [3] database is an open source spot for information on molecular interaction.

9.2.4 ChiTaRS 3.1

Chimeric Transcripts and RNA-seq [4] is a collection of chimera's PPI data along with its visualization, network, and annotation.

9.2.5 xComb

xComb [5] is a cross-linked peptide database from PPI's viewpoint.

Table 9.1 List of general PPI databases

S. No	Database	Database name	URL	References
1	DIP	Database of Interacting Proteins	http://dip.doe-mbi.ucla.edu/dip/Main.cgi	[1]
2	MINT	Molecular INTeraction Database	http://mint.bio.uniroma2.it/	[2]
3	IntAct	IntAct	http://www.ebi.ac.uk/intact/	[3]
4	ChiTaRS	Chimeric Transcripts and RNA-seq	http://chitars.md.biu.ac.il/	[4]
5	xComb	A cross-linked peptide database approach to PPI	https://goodlett.umaryland.edu/xcomb/index.php	[5]
6	ANCHOR	PPI binding pockets for drug discovery	http://structure.pitt.edu/anchor	[6]
7	ADAN	Modular domains of PPI	http://adan-embl.ibmc.umh.es/	[7]
8	2P2Idb v2	Protein-Protein Interaction Inhibition	http://2p2idb.cnrs-mrs.fr	[8]
9	Interpare	Interpare	http://interpare.net/	[9]
10	Death Domain DB	PPI for Death Domain Superfamily	http://www.deathdomain.org/	[10]
11	PDZbase	PDZbase	http://abc.med.cornell.edu/pdzbase	[11]
12	InterDom	Interacting Domains	http://interdom.i2r.a-star.edu.sg/	[12]
13	PARPs	PARPs	https://sourceforge.net/projects/parpdb/	[13]
14	PCRPi-DB	Computationally annotated hot spots in protein interfaces	http://www.bioinsilico.org/PCRPIDB	[14]

9.2.6 ANCHOR

ANCHOR [6] describes on the usage of PPI interface analysis in finding binding pockets for drug discovery.

9.2.7 ADAN

ADAN [7] reports on modular domain PPIs.

9.2.8 2P2Idb v2

2P2Idb v2 [8] serve as a resource for PDZ domain-mediated PPI.

Table 9.2 List of PPI databases for model organisms

S. No	Database	Database name	URL	References
1	MIPS	The MIPS Mammalian PPI Database	http://mips.helmholtz-muenchen.de/proj/ppi/	[15]
2	SpirPro	Spirulina-Proteome Repository	http://spirpro.sbi.kmutt.ac.th/	[16]
3	HAPPI	Human Annotated and Predicted PPI	http://bio.informatics.iupui.edu/HAPPI/	[17]
4	AraPPISite	Annotations for *Arabidopsis thaliana* PPI site	http://systbio.cau.edu.cn/arappisite/index.html	[18]
5	PPIM	A PPI Database for Maize	http://comp-sysbio.org/ppim	[19]
6	FPPI	*Fusarium graminearum* PPI database	http://csb.shu.edu.cn/fppi.	[20]
7	SPIDer	*Saccharomyces* PPI database	http://cmb.bnu.edu.cn/SPIDer/index.html.	[21]
8	PCPPI	*Penicillium*-crop protein-protein interactions	http://bdg.hfut.edu.cn/pcppi/index.html.	[22]
9	HPID	Human Protein Interaction Database	http://wilab.inha.ac.kr/hpid/	[23]
10	I2D	Interologous Interaction Database	http://ophid.utoronto.ca/ophidv2.204/index.jsp	[24]
11	HPRD	The Human Protein Reference Database	http://www.hprd.org/	[25]
12	OPHID	The Online Predicted Human Interaction Database	http://ophid.utoronto.ca/ophidv2.204/	[26]
13	PIPs	PPI prediction database	http://www.compbio.dundee.ac.uk/www-pips/	[27]
14	HitPredict	Database of High Confidence PPI	http://hintdb.hgc.jp/htp/	[28]

Table 9.3 List of network databases for PPI

S. No	Database	Database name	URL	References
1	BIND	The Biomolecular Interaction Network Database	https://www.bindingdb.org/bind/index.jsp	[29]
2	TRIP Database 2.0	TRP channel interaction network	http://www.trpchannel.org	[30]
3	STRING	STRING	https://string-db.org/	[31]
4	IntNetDB	Interaction Network Database	http://hanlab.genetics.ac.cn/IntNetDB.htm.	[32]
5	PRIMOS	Protein Interaction and Molecule Search	http://primos.fh-hagenberg.at.	[33]
6	IsoBase	IsoBase	http://isobase.csail.mit.edu/	[34]
7	ComPPI	A cellular compartment-specific database for PPI	http://ComPPI.LinkGroup.hu	[35]

Table 9.4 List of structural data repositories for PPI

S. No	Database	Database name	URL	References
1	SNAPPI-DB	Structures, iNterfaces and Alignments for PPI	http://www.compbio.dundee.ac.uk/SNAPPI/snappidb.jsp	[36]
2	SCOPPI	Structural Classification of PPI	http://scoppi.biotec.tu-dresden.de/scoppi/	[37]
3	3did	3D Interacting Domain	http://3did.irbbarcelona.org/	[38]
4	PIBASE	PIBASE	http://salilab.org/pibase	[39]
5	ProtCom	Protein complexes	http://www.ces.clemson.edu/compbio/protcom	[40]
6	JET2 Viewer	PPI sites	http://www.jet2viewer.upmc.fr/	[41]

Table 9.5 List of pathway database for PPI

S. No	Database	Database name	URL	Reference
1	KEGG	Kyoto Encyclopedia of Genes and Genomes	http://www.genome.jp/kegg/	[42]

9.2.9 Interpare

Interpare [9] has collective information on protein-domain interaction at the interface.

9.2.10 Death Domain DB

Death Domain DB [10] is a comprehensive data on death domain superfamily's PPI.

9.2.11 PDZbase

PDZbase [11] provides collective data on PDZ domains for PPI.

9.2.12 InterDom

InterDom [12] describes an integrative approach toward exploring the putative protein-domain interaction data.

9.2.13 PARPs

PARPs [13] serve as a laboratory information management system for PPI data with respect to interaction, annotation, and sequence management.

9.2.14 PCRPi-DB (Presaging Critical Residues in Protein Interfaces Database)

PCRPi-DB [14] is a public database providing a description for annotated hot spots in protein interfaces.

9.3 PPI Repositories for Model Organisms

PPI is a critical phenomenon in understanding cellular function in each species organism. Hence, it is important to study PPI using model organism. There are 14 databases describing PPI for model organism as given in Table 9.2.

9.3.1 MIPS

MIPS [15] is a mammalian protein-protein interaction database containing experimental evidence curated by experts.

9.3.2 SpirPro

SpirPro [16] discusses PPI for *Spirulina* (*Arthrospira*) and domain-domain interaction with network, visualization, and pathways.

9.3.3 HAPPI (Human Annotated and Predicted Protein Interaction)

HAPPI [17] is a repository of data on human PPI with annotation.

9.3.4 AraPPISite

AraPPISite [18] is a database for *Arabidopsis thaliana* PPI and annotation.

9.3.5 PPIM (Protein-Protein Interaction Database for Maize)

PPIM [19] discuss on maize PPI and its network.

9.3.6 FPPI (*Fusarium graminearum* Protein-Protein Interaction)

FPPI [20] provides data on *Fusarium graminearum* PPI.

9.3.7 SPIDer (*Saccharomyces* Protein-Protein Interaction Database)

SPIDer [21] is a collection of *Saccharomyces* PPI information.

9.3.8 PCPPI (*Penicillium*-Crop Protein-Protein Interactions)

PCPPI [22] is a compilation on *Penicillium* crops related to predicted PPI information.

9.3.9 HPID (Human Protein Interaction Database)

HPID [23] describes human PPI extracted using structural and experimental data of proteins.

9.3.10 I2D (Interologous Interaction Database)

I2D [24] is a model organisms PPI repository.

9.3.11 HPRD (Human Protein Reference Database)

HPRD [25] is a compendium of human protein data with updated features.

9.3.12 OPHID (Online Predicted Human Interaction Database)

OPHID [26] describes literature-driven and predicted predictions of PPI in humans.

9.3.13 PIPs (Human Protein-Protein Interaction Prediction)

PIP [27] contains predicted information on proteins to study PPI using Bayesian method that are not listed elsewhere.

9.3.14 HitPredict

HitPredict [28] is a repository on PPI in nine species with quality assessment.

9.4 PPI Network Databases

Networks are created as a result of PPI in cell and molecular biology. Therefore, it is of interest to create network maps to illustrate PPI in biological system. There are seven databases to describe PPI networks as given in Table 9.3.

9.4.1 BIND (Biomolecular Interaction Network Database)

BIND [29] provides PPI network data of biomolecules.

9.4.2 TRIP Database 2.0 (Transient Receptor Potential Channel-Interacting Protein)

TRIP [30] serves as a hub for TRP (transient receptor potential) network information.

9.4.3 STRING 7

STRING 7 [31] is an integration of predicted PPI in addition to comparative genomics, phylogenetics, and network studies.

9.4.4 IntNetDB (Integrated Network Database)

IntNetDB [32] is a consolidated repository of predicted PPI with annotations and specifically provides phenotypic distance and genetic interactions.

9.4.5 PRIMOS (Protein Interaction and Molecule Search)

PRIMOS [33] is a single database integrating information from six repositories such as BIND, DIP, HPRD, IntAct, MINT, and MIPS. These databases were discussed earlier in Section 9.2.

9.4.6 IsoBase

IsoBase [34] provides data on the functional relation existing among proteins across PPI networks of five organisms, namely, *Saccharomyces cerevisiae*, *Drosophila melanogaster*, *Caenorhabditis elegans*, *Mus musculus*, and *Homo sapiens*, that are using IsoRankN algorithm.

9.4.7 ComPPI (Compartment-Specific Database for Protein-Protein Interaction)

ComPPI [35] is an integrated database compiling network information of four species including *Saccharomyces cerevisiae*, *Drosophila melanogaster*, *Caenorhabditis elegans*, and *Homo sapiens* into one compartment to explore its biological function.

9.5 Structural Data for PPI

The use of structural data in the understanding of PPI is highly critical. Therefore, it is of interest to collect, glean, label, and archive such information in the understanding of PPI in biological systems. There are six repositories for exploring PPI using structural information as given in Table 9.4.

9.5.1 SNAPPI-DB (Structures, iNterfaces and Alignments for Protein-Protein Interactions-Database)

SNAPPI-DB [36] provides specific information on structure, interface, and alignments for each entry in the PDB (protein data bank) and also their connected information in the databases such as SCOP, CATH, Pfam, SWISSPROT, InterPro, GO terms, and protein quaternary structures (PQS).

9.5.2 SCOPPI (Structural Classification of Protein-Protein Interfaces)

SCOPPI [37] annotates and classifies domain interaction data of all known protein structures using geometric analysis of domain interfaces.

9.5.3 3did (3D Interacting Domains)

3did [38] collects information on domain-domain interaction of known 3D structure using its molecular data.

9.5.4 PIBASE

PIBASE [39] is a relational database integrating information from PDB, PQS (protein quaternary structure), CATH (class, architecture, topology, and homology), and SCOP (Structural Classification Of Proteins) to reveal structure of interfaces formed for understanding protein-protein interaction.

9.5.5 ProtCom (PROTein COMplexes)

ProtCom [40] explores all known protein structures in the PDB and simulates domain-domain structures using available data.

9.5.6 JET2 Viewer

JET2 Viewer [41] is a database for visualizing binding sites in PPI of known structures in PDB using JET2 method.

9.6 PPI in Pathways

Pathways link genes and proteins in a biochemical network. Networks describing cell division associated with several chronic diseases such as cancer are associated with PPI. The recent developments in KEGG include information related to diseases associated with PPI as given in Table 9.5.

9.6.1 KEGG (Kyoto Encyclopedia of Genes and Genomes

KEGG [42] links genes database (genomic information) and pathway database (functional data) in order to understand gene functions. They also include information associated with human diseases such as cancer, viral diseases, and others containing information related to PPI.

Exercises

1. Expand RDBMS
2. What is a database?
3. Illustrate a database schema.
4. Illustrate database architecture.
5. Illustrate the types of PPI databases using a chart.
6. What is a PPI network?
7. How structural complexes help understand PPI?
8. Expand KEGG.
9. What is the difference between domain-domain interaction (DDI) and protein-protein interaction (PPI)?
10. Illustrate the importance of database integration using an example in the understanding of PPI.

References

1. Salwinski L, Miller CS, Smith AJ, Pettit FK, Bowie JU, Eisenberg D. The database of interacting proteins. Nucleic Acids Res. 2004;32:D449–51.
2. Chatr-aryamontri A, Ceol A, Palazzi LM, Nardelli G, Schneider MV, Castagnoli L, et al. MINT: the molecular INTeraction database. Nucleic Acids Res. 2007;35:D572–4.
3. Aranda B, Achuthan P, Alam-Faruque Y, Armean I, Bridge A, Derow C, et al. The IntAct molecular interaction database in 2010. Nucleic Acids Res. 2010;38:D525–31.
4. Gorohovski A, Tagore S, Palande V, Malka A, Raviv-Shay D, Frenkel-Morgenstern M. ChiTaRS-3.1-the enhanced chimeric transcripts and RNA-seq database matched with protein-protein interactions. Nucleic Acids Res. 2017;45(D1):D790–5.
5. Panchaud A, Singh P, Shaffer SA, Goodlett DR. xComb: a cross-linked peptide database approach to protein-protein interaction analysis. J Proteome Res. 2010;9(5):2508–15.
6. Meireles LM, Dömling AS, Camacho CJ. ANCHOR: a web server and database for analysis of protein-protein interaction binding pockets for drug discovery. Nucleic Acids Res. 2010;38(Web Server issue):W407–11.
7. Encinar JA, Fernandez-Ballester G, Sánchez IE, Hurtado-Gomez E, Stricher F, Beltrao P, et al. ADAN: a database for prediction of protein-protein interaction of modular domains mediated by linear motifs. Bioinformatics. 2009;25(18):2418–24.
8. Basse MJ, Betzi S, Morelli X, Roche P. 2P2Idb v2: update of a structural database dedicated to orthosteric modulation of protein-protein interactions. Database (Oxford). 2016;pii: baw007.
9. Gong S, Park C, Choi H, Ko J, Jang I, Lee J, et al. A protein domain interaction interface database: InterPare. BMC Bioinform. 2005;6:207.
10. Kwon D, Yoon JH, Shin SY, Jang TH, Kim HG, So I, et al. A comprehensive manually curated protein-protein interaction database for the death domain superfamily. Nucleic Acids Res. 2012;40(Database issue):D331–6.
11. Beuming T, Skrabanek L, Niv MY, Mukherjee P, Weinstein H. PDZBase: a protein-protein interaction database for PDZ-domains. Bioinformatics. 2005;21(6):827–8.
12. Ng SK, Zhang Z, Tan SH, Lin K. InterDom: a database of putative interacting protein domains for validating predicted protein interactions and complexes. Nucleic Acids Res. 2003;31(1):251–4.
13. Droit A, Hunter JM, Rouleau M, Ethier C, Picard-Cloutier A, Bourgais D, et al. PARPs database: a LIMS systems for protein-protein interaction data mining or laboratory information management system. BMC Bioinform. 2007;8:483.
14. Segura J, Fernandez-Fuentes N. PCRPi-DB: a database of computationally annotated hot spots in protein interfaces. Nucleic Acids Res. 2011;39(Database issue):D755–60.
15. Pagel P, Kovac S, Oesterheld M, Brauner B, Dunger-Kaltenbach I, Frishman G, et al. The MIPS mammalian protein protein interaction database. Bioinformatics. 2005;21:832–4.
16. Senachak J, Cheevadhanarak S, Hongsthong A. SpirPro: a spirulina proteome database and web-based tools for the analysis of protein-protein interactions at the metabolic level in Spirulina (Arthrospira) platensis C1. BMC Bioinform. 2015;16:233.
17. Chen JY, Mamidipalli S, Huan T. HAPPI: an online database of comprehensive human annotated and predicted protein interactions. BMC Genomics. 2009;10(Suppl 1):S16.
18. Li H, Yang S, Wang C, Zhou Y, Zhang Z. AraPPISite: a database of fine-grained protein-protein interaction site annotations for Arabidopsis Thaliana. Plant Mol Biol. 2016;92(1–2):105–16.
19. Zhu G, Wu A, XJ X, Xiao PP, Lu L, Liu J, et al. PPIM: a protein-protein interaction database for maize. Plant Physiol. 2016;170(2):618–26.
20. Zhao XM, Zhang XW, Tang WH, Chen L. FPPI: fusarium graminearum protein-protein interaction database. J Proteome Res. 2009;8(10):4714–21.
21. Wu X, Zhu L, Guo J, Fu C, Zhou H, Dong D *et al.* SPIDer: Saccharomyces protein-protein interaction database. BMC Bioinform. 2006;7(Suppl 5):S16.
22. Yue J, Zhang D, Ban R, Ma X, Chen D, Li G. PCPPI: a comprehensive database for the prediction of Penicillium-crop protein-protein interactions. Database (Oxford). 2017;2017(1)

23. Han K, Park B, Kim H, Hong J, Park J. HPID: the human protein interaction database. Bioinformatics. 2004;20(15):2466–70.
24. Brown KR, Jurisica I. Unequal evolutionary conservation of human protein interactions in interologous networks. Genome Biol. 2007;8:R95.
25. Prasad TSK, Goel R, Kandasamy K, Keerthikumar S, Kumar S, Mathivanan S, et al. Human protein reference database. Nucleic Acids Res. 2009;37:D767–72.
26. Brown KR, Jurisica I. Online predicted human interaction database. Bioinformatics. 2005;21(9):2076–82.
27. McDowall MD, Scott MS, Barton GJ. PIPs: human protein-protein interaction prediction database. Nucleic Acids Res. 2009;37(Database issue):D651–6.
28. Patil A, Nakai K, Nakamura H. HitPredict: a database of quality assessed protein-protein interactions in nine species. Nucleic Acids Res. 2011;39(Database issue):D744–9.
29. Isserlin R, El-Badrawi RA, Bader GD. The biomolecular interaction network database in PSI-MI 2.5. Database (Oxford). 2011;2011:baq037.
30. Shin YC, Shin SY, Chun JN, Cho HS, Lim JM, Kim HG *et al.* TRIP database 2.0: a manually curated information hub for accessing TRP channel interaction network. PLoS One. 2012;7(10):e47165.
31. Von Mering C, Jensen LJ, Kuhn M, Chaffron S, Doerks T, Kruger B, et al. STRING 7 – recent developments in the integration and prediction protein networks. Nucleic Acids Res. 2007;35:D358–62.
32. Xia K, Dong D, Han JD. IntNetDB v1.0: an integrated protein-protein interaction network database generated by a probabilistic model. BMC Bioinform. 2006;7:508.
33. Rid R, Strasser W, Siegl D, Frech C, Kommenda M, Kern T, et al. PRIMOS: an integrated database of reassessed protein-protein interactions providing web-based access to in silico validation of experimentally derived data. Assay Drug Dev Technol. 2013;11(5):333–46.
34. Park D, Singh R, Baym M, Liao CS, Berger B. IsoBase: a database of functionally related proteins across PPI networks. Nucleic Acids Res. 2011;39(Database issue):D295–300.
35. Veres DV, Gyurkó DM, Thaler B, Szalay KZ, Fazekas D, Korcsmáros T *et al.* ComPPI: a cellular compartment-specific database for protein-protein interaction network analysis. Nucleic Acids Res. 2015;43(Database issue):D485–93. network.
36. Jefferson ER, Walsh TP, Roberts TJ, Barton GJ. SNAPPI-DB: a database and API of structures, iNterfaces and alignments for protein-protein interactions. Nucleic Acids Res. 2007;35(Database issue):D580–9.
37. Winter C, Henschel A, Kim WK, Schroeder M. SCOPPI: a structural classification of protein-protein interfaces. Nucleic Acids Res. 2006;34(Database issue):D310–4.
38. Mosca R, Céol A, Stein A, Olivella R, Aloy P. 3did: a catalog of domain-based interactions of known three-dimensional structure. Nucleic Acids Res. 2014;42(Database issue):D374–9.
39. Davis FP, Sali A. PIBASE: a comprehensive database of structurally defined protein interfaces. Bioinformatics. 2005;21(9):1901–7.
40. Kundrotas PJ, Alexov E. PROTCOM: searchable database of protein complexes enhanced with domain-domain structures. Nucleic Acids Res. 2007;35(Database issue):D575–9.
41. Ripoche H, Laine E, Ceres N, Carbone A. JET2 viewer: a database of predicted multiple, possibly overlapping, protein-protein interaction sites for PDB structures. Nucleic Acids Res. 2017;45(D1):D236–42.
42. Kanehisa M, Goto S. KEGG: kyoto encyclopedia of genes and genomes. Nucleic Acids Res. 2000;28:27–30.

Homodimer Protein Folding and Binding 10

10.1 Homodimer Folding Mechanisms

Homodimer protein folding is studied by measuring the denatured species fraction using spectroscopic techniques such as fluorescence, circular dichroism (CD), nuclear magnetic resonance (NMR), and absorption. It is known that homodimer protein folding occurs through 2-state (2S) and 3-state (3S). The 3S folding occurs through either a monomer intermediate (3SMI) or dimer intermediate 3SDI as shown in Fig. 10.1. A number of homodimer protein structures are known with folding information as summarized in Table 10.1 and as described elsewhere [1–4]. The structural features describing homodimer interface in relation to known folding mechanism are now available in

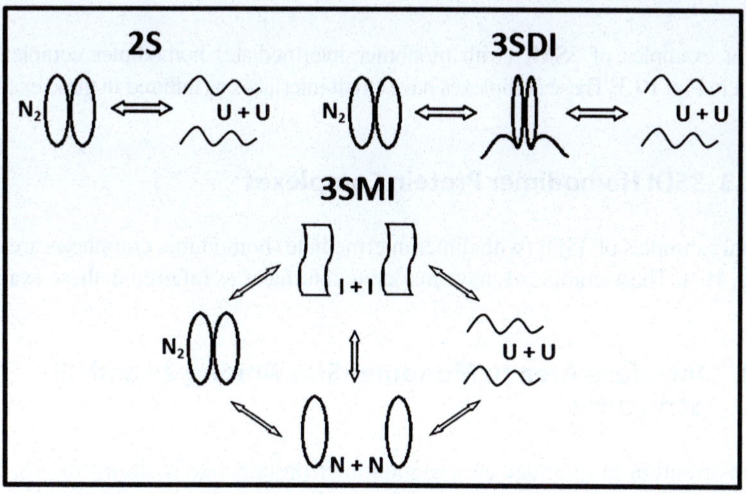

Fig. 10.1 An illustration of 2-state and 3-state homodimer protein-protein complex folding and binding. *2S* 2-state, *3S* 3-state, *3SMI* 3-state monomer intermediate, *3SDI* 3-state dimer intermediate, N_2 native dimer structure, *U* unfolded, *I* intermediate, *N* native monomer structure

P. Kangueane, C. Nilofer, *Protein-Protein and Domain-Domain Interactions*,
https://doi.org/10.1007/978-981-10-7347-2_10

Table 10.1 Datasets of *homodimer* structures with known folding mechanism

Folding mechanism	Homodimer structures (PDB ID)
2S (25)	1A7G, 1A8G, 1ARR, 1B8Z, 1BET, 1BFM, 1BUO, 1CMB, 1EE1, 1ETY, 1GSD, 1GTA, 1HTI, 1OH0, 1ROP, 1SIV, 1VQB, 1VUB, 1WRP, 1Y7Q, 2BQP, 2CPG, 2GSR, 3SSI, 5CRO
3SMI (10)	1A43, 1DFX, 1LGS, 1ND5, 1QLL, 1RUN, 1SPD, 1TYA, 1YAI, 2CRK
3SDI (6)	1AOZ, 1CM7, 1HQO, 1MUL, 1NL3, 1PSC

detail [1–4]. Table 10.1 shows a list of PDB structures grouped into 2S (Fig. 10.2), 3SMI (Fig. 10.3), and 3SDI (Fig. 10.4) based on the folding mechanism of homodimer protein complexes. Table 10.2 gives protein function-related information associated with 2S homodimers, while Table 10.3 documents for 3SMI and Table 10.4 for 3SDI. The structures in Figs. 10.2, 10.3, and 10.4 provide valuable insights to qualitatively relate homodimer folding mechanism with known molecular function.

10.2 Homodimer Structures

10.2.1 2S Homodimer Protein Complexes

Several examples of 2S (without intermediate) homodimer complexes are shown in Fig. 10.2. These complexes have large interfaces as inferred in these examples.

10.2.2 3SMI Homodimer Protein Complexes

Several examples of 3SMI (with monomer intermediate) homodimer complexes are shown in Fig. 10.3. These complexes have small interfaces as inferred in these examples.

10.2.3 3SDI Homodimer Protein Complexes

Several examples of 3SDI (with dimer intermediate) homodimer complexes are shown in Fig. 10.4. These complexes have moderate interfaces as inferred in these examples.

10.3 Interface Area to Monomer Size Among 2S and 3S Structures

The distribution of interface area to monomer protein size is shown for 2S, 3SMI, and 3SDI homodimer protein complexes in Fig. 10.5. The distribution shows that 2S homodimer complexes are relatively small in size with large interface area. It is also clear that 3SDI homodimer complexes have moderate interfaces, while 3SMI do have small interfaces in several cases. Thus, the relationship between monomer protein size and interface area among 2S, 3SMI, and 3SDI homodimer protein complexes is demonstrated.

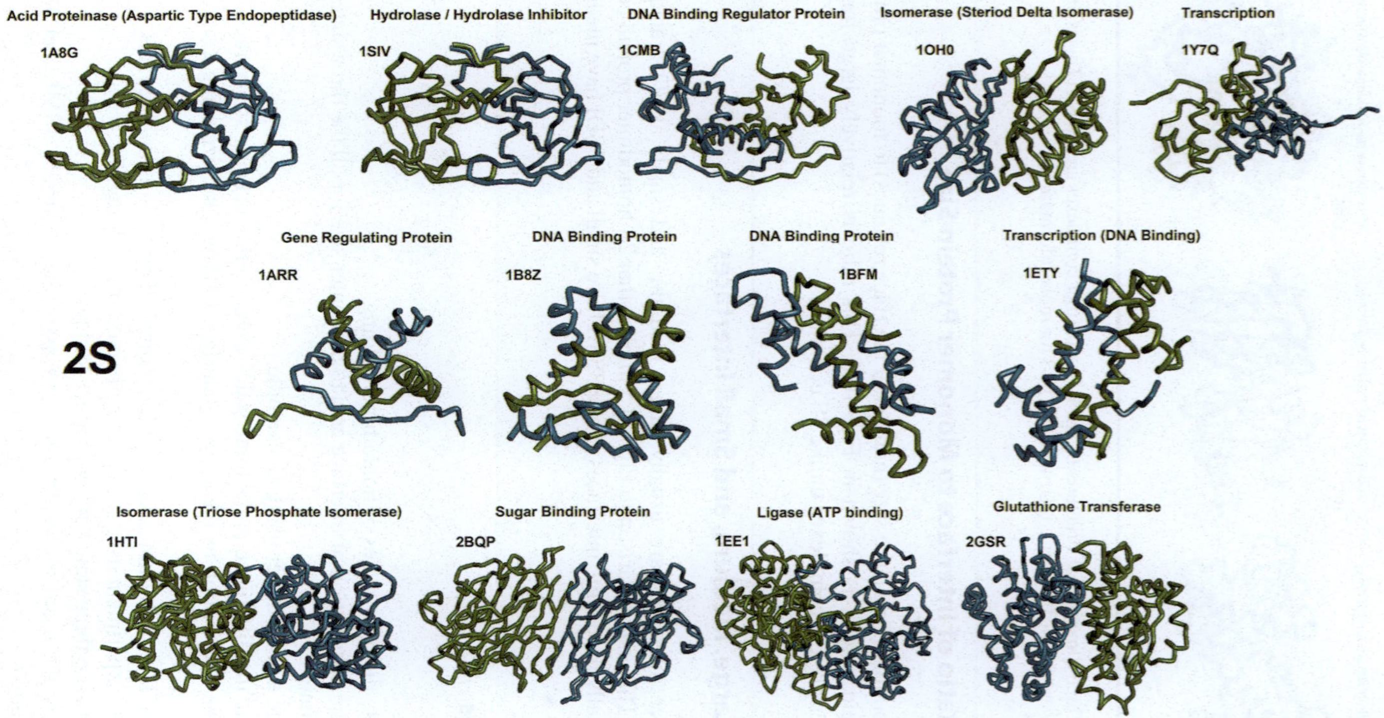

Fig. 10.2 Three-dimensional structures of 2S homodimer complexes are shown (This is drawn using Discovery Studio™. Please refer to Table 10.2 for more information)

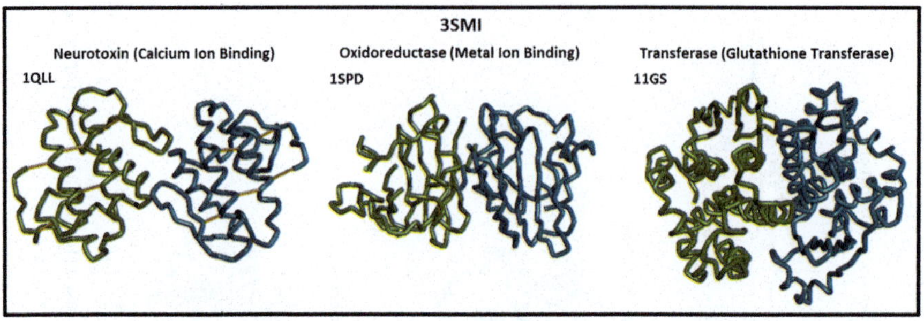

Fig. 10.3 Three-dimensional structures of 3SMI homodimer complexes are shown (This is drawn using Discovery Studio™. Please refer to Table 10.3 for more information)

10.4 Ratio of Interface to Monomer Protein Size

The distribution of interface size (number of interface residues) to monomer protein size (protein length) is shown in Fig. 10.6. This ratio is relatively large for 2S homodimer protein complexes unlike 3S complexes.

10.5 Large, Medium, and Small Interfaces

Homodimer interfaces are usually large, medium, and small as illustrated in Fig. 10.7. This in comparison to Fig. 10.5 implies that 2S homodimer protein complexes usually have large interfaces, while 3SMI have small and 3SDI have medium interfaces.

Exercises

1. Illustrate the three states of homodimer folding.
2. Show the distribution of interface area to monomer size for all the three states of homodimers.
3. Illustrate a homodimer using an example.
4. Illustrate the difference in the three states of homodimers using structures.
5. Show the distribution of the ratio of interface to monomer size using a suitable chart.
6. Illustrate large, medium, and small homodimer interfaces using diagrams.
7. State the experiment used for studying homodimer folding.
8. What are the three states of homodimer folding?
9. What are the usual molecular functions associated with 2S, 3SMI, and 3SDI homodimers?

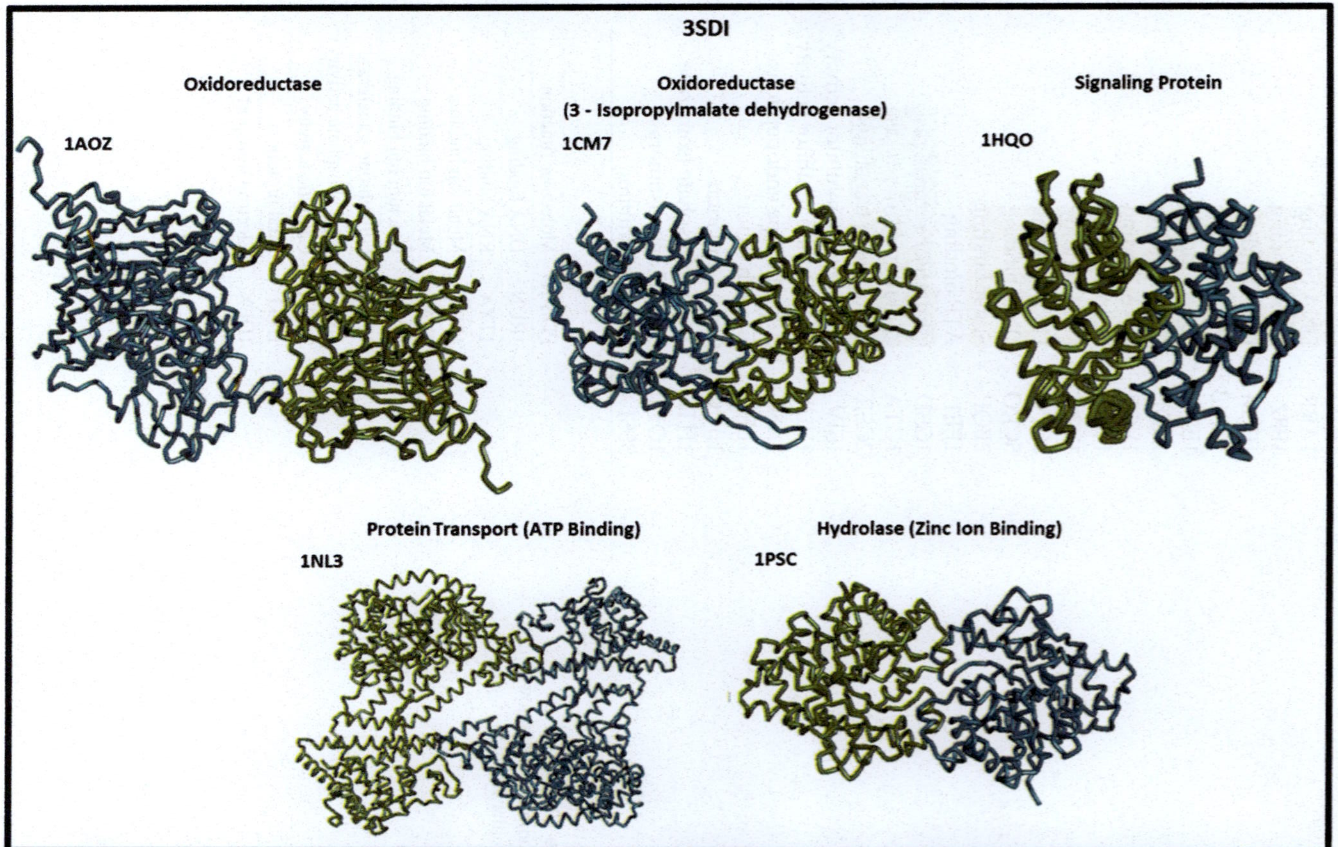

Fig. 10.4 Three-dimensional structures of 3SDI homodimer complexes are shown (This is drawn using Discovery Studio™. Please refer to Table 10.4 for more information)

Table 10.2 2S homodimers with known molecular functions

PDB ID	Molecular Function
1A7G	DNA binding
1ARR	DNA binding
1B8Z	DNA binding
1BFM	DNA binding
1CMB	DNA binding
1ETY	DNA binding
1ROP	DNA binding
1VQB	DNA binding
1WRP	DNA binding
1Y7Q	DNA binding
2CPG	DNA binding
5CRO	DNA binding
2BQP	Sugar binding
1EE1	ATP binding
1GSD	Glutathione transferase
1GTA	Glutathione transferase
2GSR	Glutathione transferase
1SIV	DNA gyrase inhibitor activity
1VUB	DNA gyrase inhibitor activity
1A8G	Aspartic type endopeptidase
1BET	Growth factor
1BUO	Gene regulation
1HTI	Triose phosphate isomerase
1OH0	Steriod delta isomerase
3SSI	Enzyme inhibitor

Table 10.3 3SMI homodimers with known molecular function

PDB ID	Molecular function
1RUN	DNA binding
1TYA	RNA binding
1SPD	Metal ion binding
1YAI	Metal ion binding
1QLL	Calcium ion binding
1LGS	Glutathione transferase
1ND5	Acid phosphate activity
2CRK	Catalytic activity
1A43	Viral protein
1DFX	Electron transport

Table 10.4 3SDI homodimers with known molecular functions

PDB ID	Molecular function
1MUL	DNA binding
1NL3	ATP binding
1PSC	Zinc ion binding
1AOZ	Oxidoreductase activity
1CM7	3-Isopropylmalate dehydrogenase activity
1HQO	Signaling protein

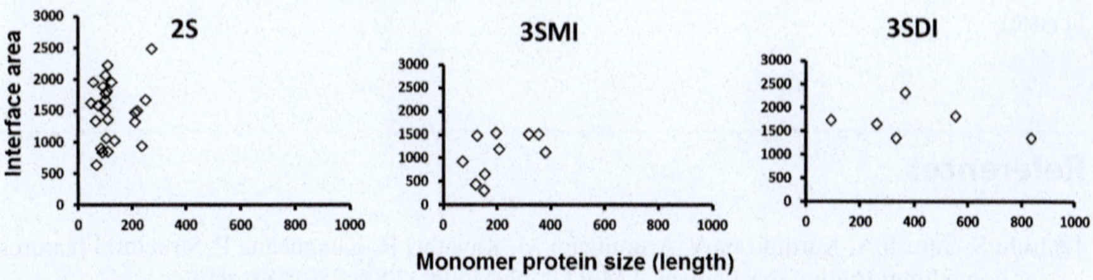

Fig. 10.5 The distribution of 2S, 3SMI, and 3SDI homodimer complexes based on interface area and monomer protein size. 2S has large size interfaces, 3SMI shows small size interfaces, and 3SDI has medium size interfaces (This is adapted from Karthikraja et al. [3] under the open access creative commons attribution license)

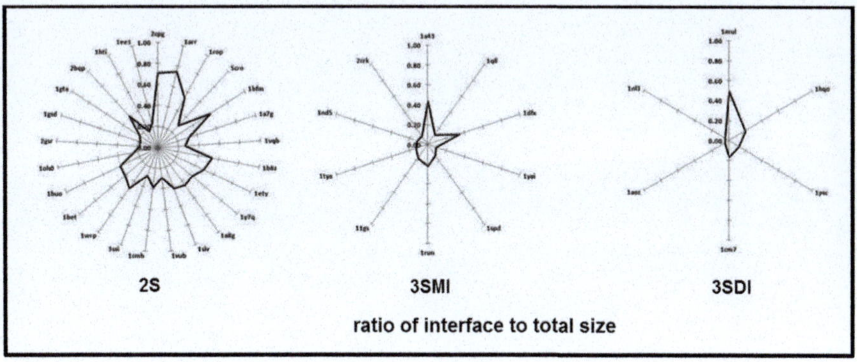

Fig. 10.6 The distribution of ratio of interface to total size among 2S, 3SMI, and 3SDI is shown. This ratio is relatively large for 2S compared to 3SMI and 3SDI (This is adapted from Karthikraja et al. [3] under the open access creative commons attribution license)

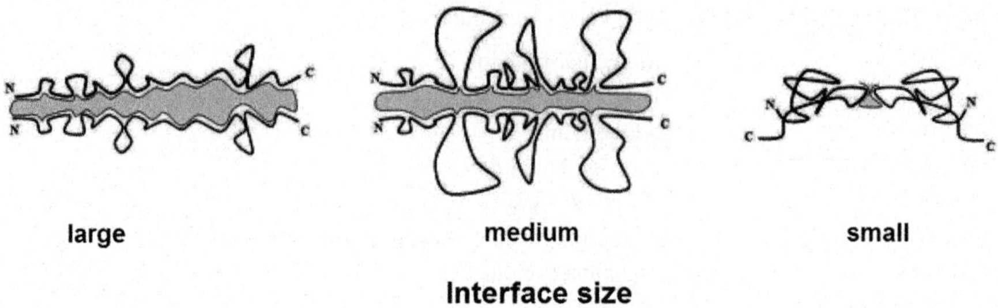

Fig. 10.7 An illustration of large, medium, and small interfaces is shown among homodimers (This is adapted from Karthikraja et al. [3] under the open access creative commons attribution license)

References

1. Lulu S, Suresh A, Karthikraja V, Arumugam M, Kayathri R, Kangueane P. Structural features for homodimer folding mechanism. J Mol Graph Model. 2009;28(2):88–94.
2. Suresh A, Karthikraja V, Lulu S, Kangueane U, Kangueane P. A decision tree model for the prediction of homodimer folding mechanism. Bioinformation. 2009;4(5):197–205.
3. Karthikraja V, Suresh A, Lulu S, Kangueane U, Kangueane P. Types of interfaces for homodimer folding and binding. Bioinformation. 2009;4(3):101–11.
4. Li L, et al. Structural features differentiate the mechanisms between 2S (2 state) and 3S (3 state) folding homodimers. Bioinformation. 2005;1(2):42–9.

Protein Subunit-Subunit to Domain-Domain Interactions

11

11.1 Fusion Protein or Rosetta Stone

Rosetta stone is described as a naturally occurring fusion protein of two or more component proteins as shown in Fig. 11.1. This occurs through an important evolutionary phenomenon called gene fusion. They try to be like operons, create protein-protein interfaces, and show multiple functions and alternative splicing across diverse evolutionary distances [1]. Fusion proteins mimicking protein-protein interface as shown in Fig. 11.1 are of great interest. Two subunits form a functional interface in one species organism (species 2), and the same interface is made by a domain-domain interface in a fused protein in yet another species organism (species 1) as shown in Fig. 11.1. We describe this phenomenon using some rare examples with known structures as shown in Figs. 11.2, 11.3, 11.4, 11.5, 11.6, and 11.7.

11.2 HisH-HisF Interface

The imidazole glycerol phosphate synthetase (IGPS) consists of glutaminase unit (HisF) and cyclase unit (HisH). The HisF-HisH domains are linked forming a rosetta stone in *S. cerevisiae* (Fig. 11.2). The functional interface is formed between the HisH and HisF domains. However, these two units are unfused in *T. thermophilus*, and the functional interface is between the two subunits. The IGPS proteins are homologous in these two species. Nonetheless, their functional interface state is different in these two species. It is a domain-domain interface in *S. cerevisiae* and a subunit-subunit interface in *T. thermophilus*, while their molecular function is analogous in nature. The state of interface is different in both the cases. Figure 11.2 shows that interface area, gap volume, and radius of gyration (Rg) are larger for HisH-HisF in *S. cerevisiae* over a molecular dynamics simulation period of 5 ns compared to its counterpart in *T. thermophilus*. Nevertheless, gap index remains the same in both the cases as shown in Fig. 11.2. This is interesting. This observation provides valuable insights to understand the formation of fusion proteins in

P. Kangueane, C. Nilofer, *Protein-Protein and Domain-Domain Interactions*,
https://doi.org/10.1007/978-981-10-7347-2_11

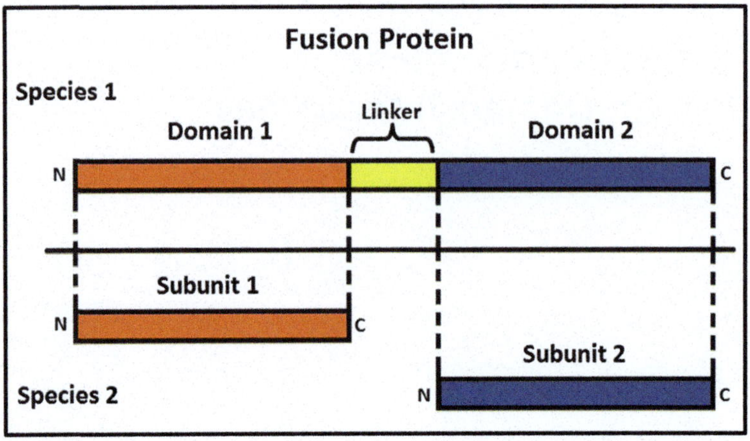

Fig. 11.1 A “rosetta stone” or “fusion protein” is illustrated using a neatly labeled diagram. A subunit-subunit protein complex exhibiting a protein-protein interaction in one species organism is found to be in a “fused form” with a linker stretch forming a fusion protein consisting of two domains in yet another species organism. This occurs in evolution by gene fusion. It should be noted that subunit 1 and domain 1 are homologous. Similarly, subunit 2 is homologous to domain 2 in this “rosetta stone” structure

molecular evolution. This infers that the fused protein architecture with greater interface area, gap volume, and Rg with conversed gap index is preferred over its unfused counterparts. Thus, chimeric, fused, rosetta stone architecture is preferred in evolution probably driven by biochemical pressure in biological systems.

11.3 DNA Gyrase A and B Interface

The topoisomerase II fusion protein in *S. cerevisiae* is made of DNA gyrase subunit B and DNA gyrase subunit A in *E. coli* as shown in Fig. 11.3. The DNA gyrase subunit B is 44% similar to the chimeric protein at the N terminus and 41% similar to the chimeric protein at the C terminus. The A and B domains in yeast are connected by a short linker as shown in Fig. 11.3. Available structures for the dimer DNA gyrase subunits A and B (PDB ID: 3NUH) and topoisomerase II fusion protein (PDB ID: 1BJT) are shown. Thus, rosetta stonelike fusion proteins with known structures for both fused and unfused forms of topoisomerase II are illustrated.

11.4 Acetate CoA Transferase α and β Interface

The succinyl CoA transferase fusion protein in human is a fusion of acetate CoA transferase α and β subunits in *E. coli* as shown in Fig. 11.4. The alpha subunit is 62% similar to the fusion protein at the N terminal, while beta subunit is 66% similar to the fusion protein at the C terminal. The subunit-subunit (PDB ID: 5DBN)

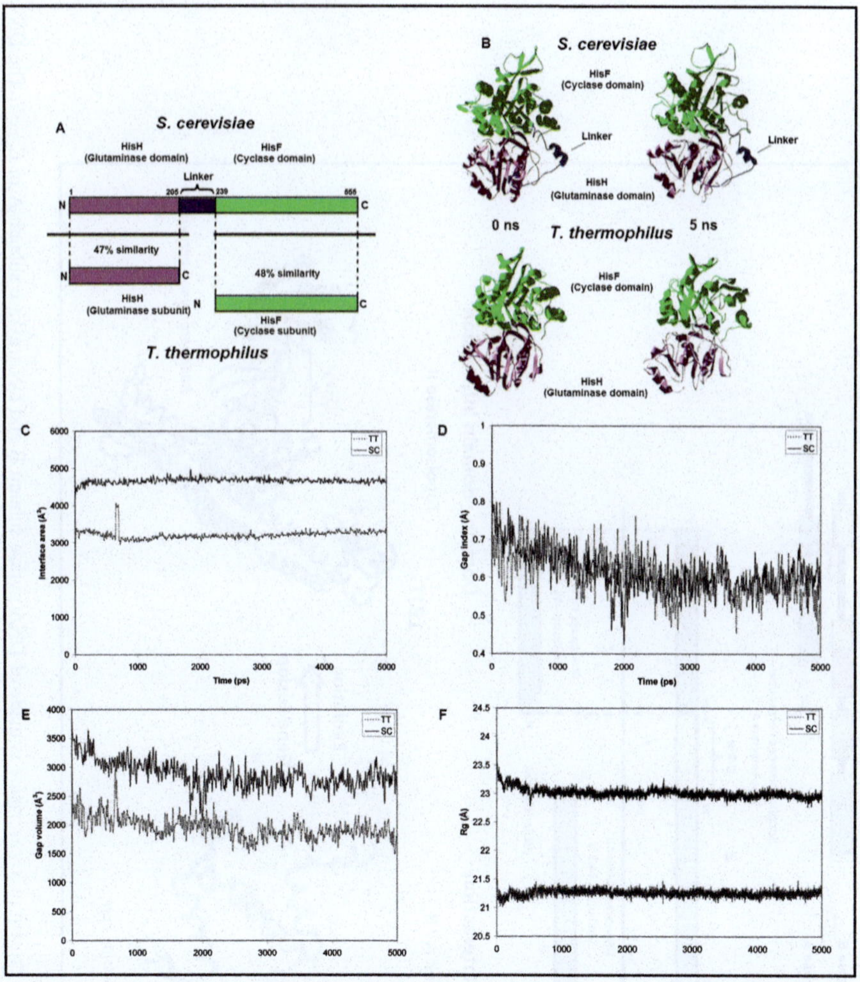

Fig. 11.2 **(a)** The fusion protein of HisF-HisH is illustrated. The HisH is homologous at 47% sequence similarity between *T. thermophilus* (TT) and *S. cerevisiae* (SC). Similarly, the HisF is homologous at 48% sequence similarity between TT and SC. However, it should be noted that HisF and HisH are connected by a linker in SC unlike in TT where HisF and HisH are subunits bound together using nonbonded interactions. **(b)** Structures of HisF-HisH fusion protein in SC and its corresponding HisF-HisH subunits complex. **(c)** The interface area profile for HisF-HisH in SC and TT over 5 ns molecular dynamics simulation is shown. **(d)** The gap volume profile for HisF-HisH in SC and TT over 5 ns molecular dynamics simulation is shown. **(e)** The gap index profile for HisF-HisH in SC and TT over 5 ns molecular dynamics simulation is shown. **(f)** The radius of gyration (R_g) profile for HisF-HisH in SC and TT over 5 ns molecular dynamics simulation is shown. Structures are known in both dimer form and fusion form for HisF-HisH (This is adapted from Yiting et al. [2] under the open access creative commons attribution license)

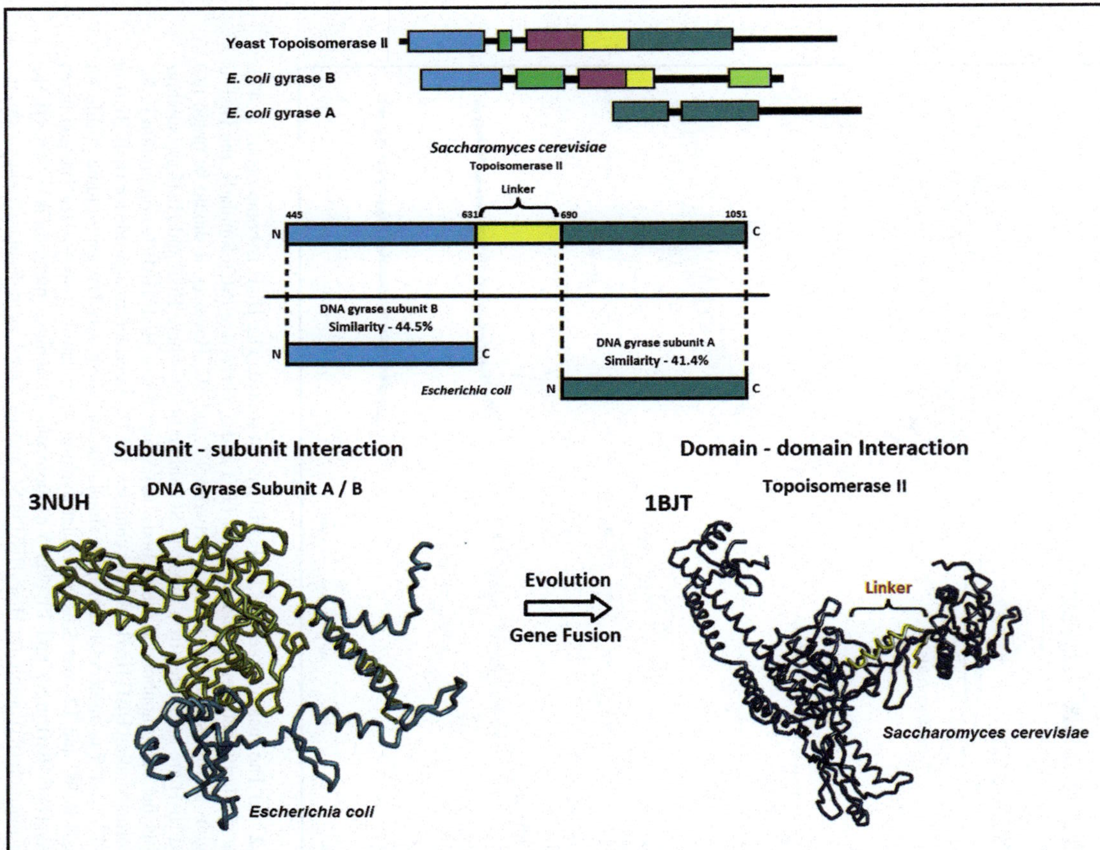

Fig. 11.3 The chimeric topoisomerase II fusion protein in *S. cerevisiae* is made of DNA gyrase subunit B and DNA gyrase subunit A in *E. coli*. The DNA gyrase subunit B is 44% similar to the chimeric protein at the N terminus and 41% similar to the chimeric protein at the C terminus. The A and B domains in yeast are connected by a short linker as shown above. Available structures for the dimer DNA gyrase subunits A and B (PDB ID: 3NUH) and topoisomerase II fusion protein (PDB ID: 1BJT) are shown

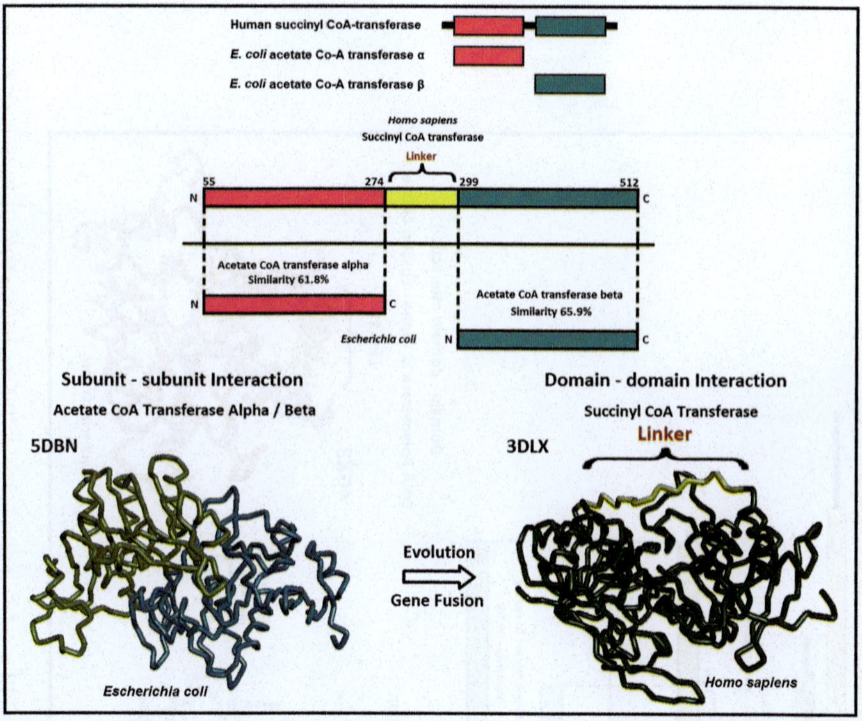

Fig. 11.4 The succinyl CoA transferase fusion protein in human is a fusion of acetate CoA transferase α and β subunits in *E. coli*. The alpha subunit is 62% similar to the fusion protein at the N terminal, while beta subunit is 66% similar to the fusion protein at the C terminal. The subunit-subunit (PDB ID: 5DBN) dimer interaction in *E. coli* and domain-domain interaction (PDB ID: 3DLX) in the fused human protein are shown using known structures

dimer interaction in *E. coli* and domain-domain interaction (PDB ID: 3DLX) in the fused human protein are shown using known structures.

11.5 DNA pol III α and ε Interface

The DNA pol III α fusion protein in *B. subtilis* is a fusion of DNA pol III α and ε subunits in *E. coli* as shown in Fig. 11.5. The α subunit is 99% similar at the N terminal of the chimeric protein, while the ε subunit is 31% similar at the C terminal of the fusion protein. The fusion protein structure (PDB ID: 4GXB) is shown. It should be noted that monomer structures are known for DNA pol III α (PDB ID: 2HNH) and ε (PDB ID: 2IDO) proteins. However, structures for fusing components in dimer form are not available despite their biochemical binding information.

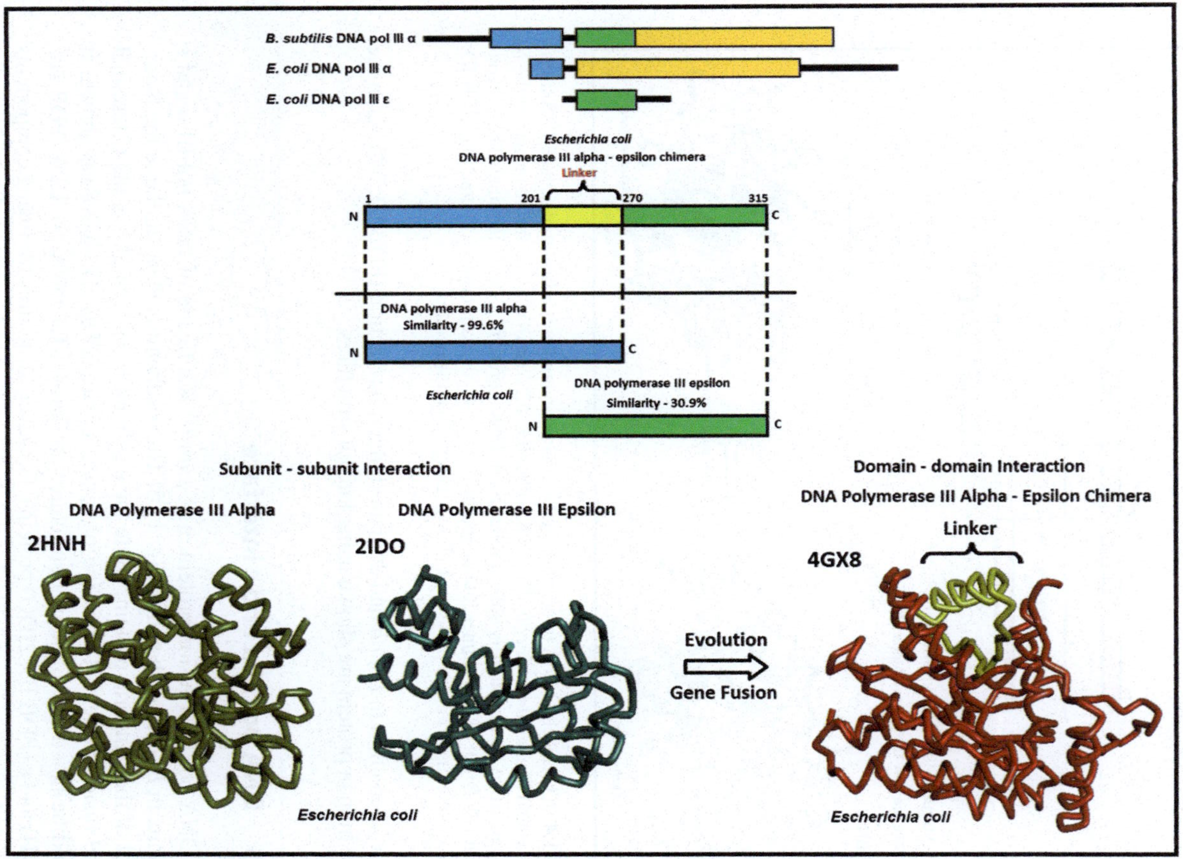

Fig. 11.5 The DNA pol III α fusion protein in *B. subtilis* is a fusion of DNA pol III α and ε subunits in *E. coli*. The α subunit is 99% similar at the N terminal of the chimeric protein, while the ε subunit is 31% similar at the C terminal of the fusion protein. The fusion protein structure (PDB ID: 4GXB) is shown. It should be noted that monomer structures are known for DNA pol III α (PDB ID: 2HNH) and ε (PDB ID: 2IDO) proteins

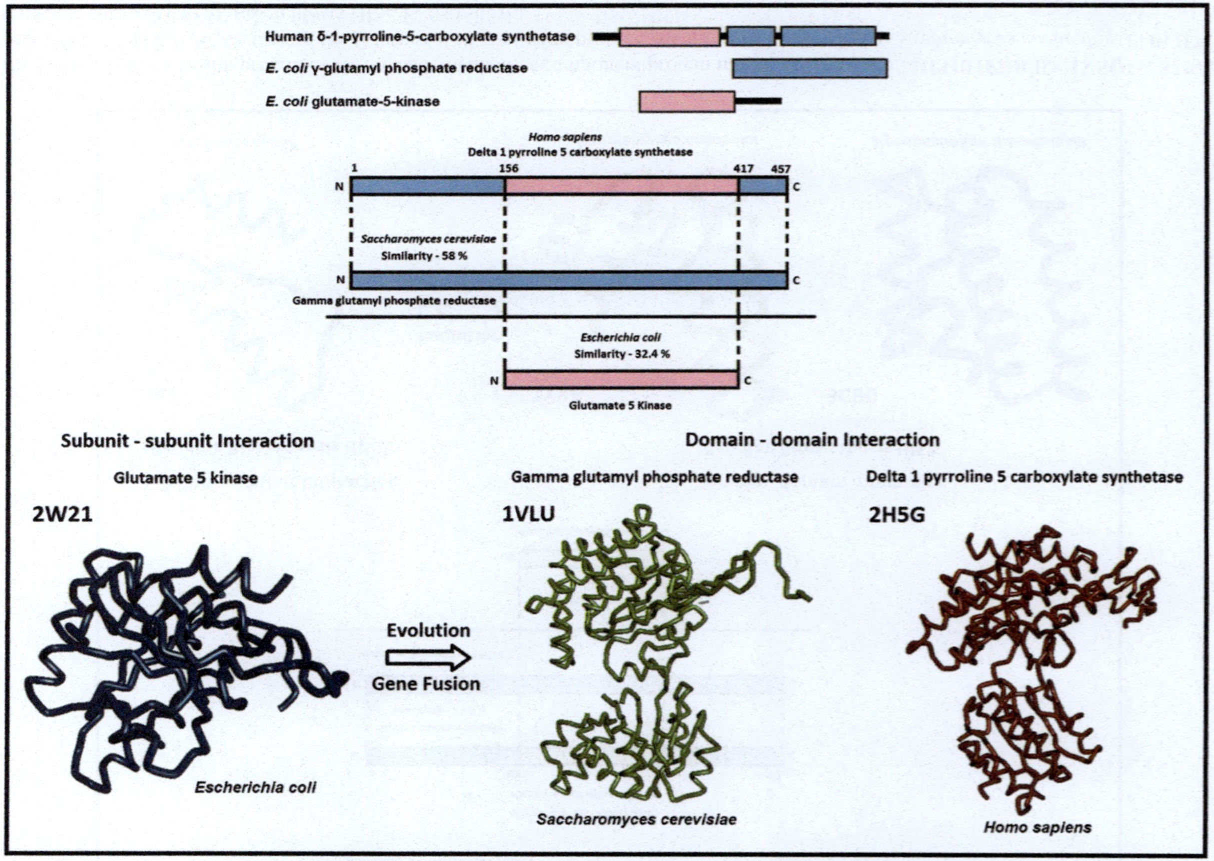

Fig. 11.6 The human δ-1-pyrroline-5-carboxylate synthetase is a fusion protein of γ-glutamyl phosphate reductase and glutamate 5-kinase in *E. coli* as shown. Structures are known for human δ-1-pyrroline-5-carboxylate synthetase (PDB ID: 2H5G) and *E. coli* glutamate 5-kinase (PDB ID: 2W21) as shown above. The glutamate 5-kinase is about 32% similar to human δ-1-pyrroline-5-carboxylate synthetase at the N terminal of the fusion protein. It should be noted that a similar fusion protein (PDB ID: 1VLU) in *S. cerevisiae* is 58% similar to the human δ-1-pyrroline-5-carboxylate synthetase as shown above

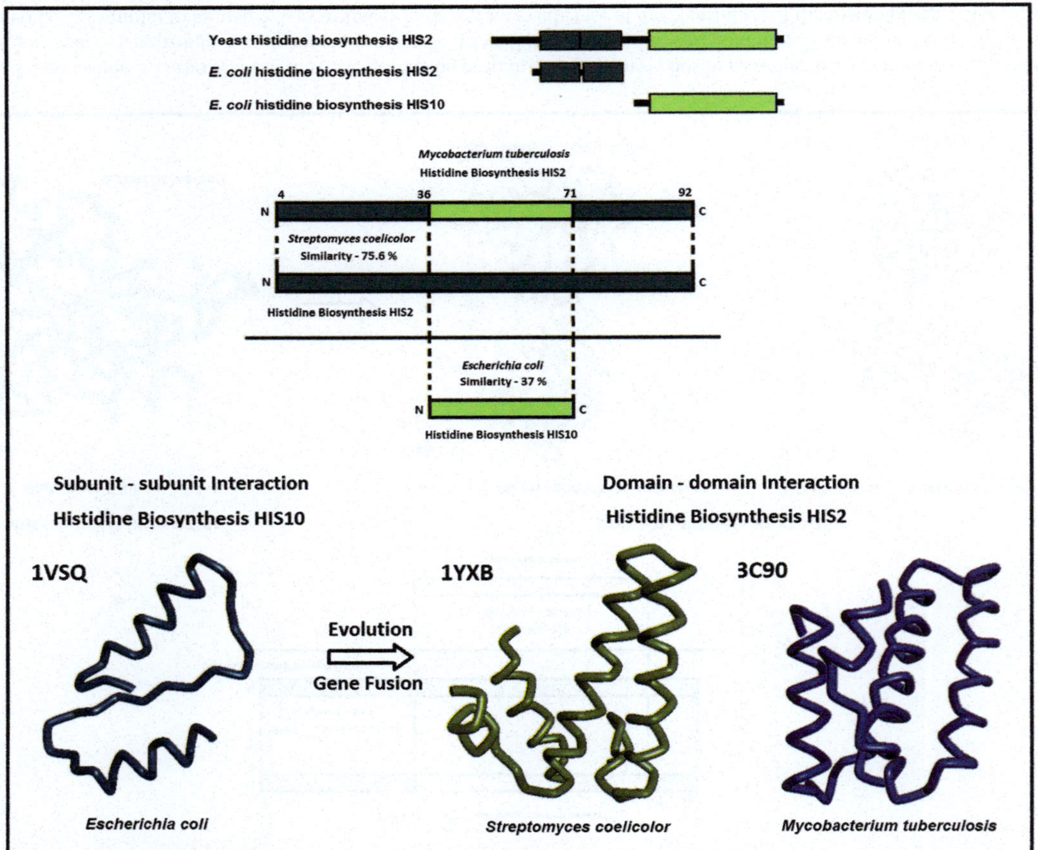

Fig. 11.7 The yeast HIS2 is a fusion protein of HIS10 and yet another component protein in *E. coli*. The *E. coli* HIS10 (PDB ID: 1VSQ) is 37% similar to the *M. tuberculosis* HIS2 (PDB ID: 3C90) toward the C terminal of the fusion protein. A similar fusion protein in *Streptomyces coelicolor* (PDB ID: 1YXB) with about 76% sequence similarity to *M. tuberculosis* HIS2 is also shown

11.6 Glutamyl Phosphate Reductase-Glutamate Kinase Interface

The human δ-1-pyrroline-5-carboxylate synthetase is a fusion protein of γ-glutamyl phosphate reductase and glutamate 5-kinase in *E. coli* as shown in Fig. 11.6. Structures are known for human δ-1-pyrroline-5-carboxylate synthetase (PDB ID: 2H5G) and *E. coli* glutamate 5-kinase (PDB ID: 2W21) as shown above. The glutamate 5-kinase is about 32% similar to human δ-1-pyrroline-5-carboxylate synthetase at the N terminal of the fusion protein. It should be noted that a similar fusion protein (PDB ID: 1VLU) in *S. cerevisiae* is 58% similar to the human δ-1-pyrroline-5-carboxylate synthetase as shown.

11.7 His2-His10 Interface

The yeast HIS2 is a fusion protein of HIS10, and yet another component protein in *E. coli* is shown in Fig. 11.7. The *E. coli* HIS10 (PDB ID: 1VSQ) is 37% similar to the *M. tuberculosis* HIS2 (PDB ID: 3C90) toward the C terminal of the fusion protein. A similar fusion protein in *Streptomyces coelicolor* (PDB ID: 1YXB) with about 76% sequence similarity to *M. tuberculosis* HIS2 is also shown.

11.8 Conclusion

A number of fusion proteins or chimeric proteins or rosetta stone are observed in recent years [1, 2]. A fusion protein observed in one species organism is made of two or more fusing components in other species organisms toward improved functionality in biological networks [3]. This is described using an important evolutionary phenomenon called gene fusion. In this context, we described the fusion structures mimicking interfaces of HisH-HisF, DNA gyrase A and B, acetate CoA transferase α and β, DNA pol III α and ε, glutamyl phosphate reductase and glutamate kinase, and His2-His10 subunits. We foresee more clarity in future with additional data with known structures for fusion proteins.

Exercises

1. What is a chimera protein?
2. What is a rosetta stone?
3. What is a fusion protein?
4. Show the HisF-HisH interface in *T. thermophilus* (TT) and *S. cerevisiae* (SC) using a neat diagram.
5. What is gene fusion?
6. Illustrate the structural evidence for fusion proteins using examples

7. Enumerate the mechanism of gene fusion.
8. Describe the hypothesis of gene fusion.
9. Illustrate the evolutionary pressure for gene fusion using examples.
10. What is gap index?
11. Give an account of fusion proteins using known structures.
12. Illustrate the DNA gyrase A and B interface with its fusion counterpart.
13. Illustrate the acetate CoA transferase α and β interface with its fusion counterpart.
14. Illustrate the DNA pol III α and ε interface with its fusion counterpart.
15. Illustrate the glutamyl phosphate reductase and glutamate kinase interface with its fusion counterpart.
16. Illustrate the His2-His10 interface with its fusion counterpart.

References

1. Yiting Y, Chaturvedi I, Meow LK, Kangueane P, Sakharkar MK. Can ends justify the means? Digging deep for human fusion genes of prokaryotic origin. Front Biosci 2004; 9:2964–2971. PubMed PMID: 15353329.
2. Yiting Y, Lei L, Sakharkar MK, Kangueane P. Insight into gene fusion from molecular dynamics simulation of fused and un-fused IGPS (imidazole glycerol phosphate Synthetase). Bioinformation 2006;1 (3):99–104. PubMed PMID: 17611615; PubMed Central PMCID: PMC1904513.
3. Sakharkar MK, Yiting Y, Chow VT, Kangueane P. Insights to metabolic network evolution by fusion proteins. Front Biosci 2005; 10:1070–1078. PubMed PMID: 15769606.

Domain-Domain Interactions 12

12.1 Chimeric Protein with Dimer-Dimer Interface

Several examples of multi-domain chimeric (fusion) proteins are available with known three-dimensional (3D) structural information as shown in Fig. 12.1. The number of such multi-domain proteins exhibiting domain-domain interface is increasingly high from lower to higher species organism. Therefore, it is of interest to systematically document such phenomenon for a better understanding of protein structure evolution towards complex biochemical function and molecular regulation.

12.2 Domain-Domain Interaction (DDI) Databases

The amount of domain information in multi-domain proteins is increasing day by day through advancement in analytical techniques for a better understanding of this phenomenon. These information are now archived in specialized databases such as IDDI [1], an integrated system; DOMINE [2], a comprehensive collection system; 3did Update [3] of structural domains; and PROTCOM [4] with enhanced DDI.

12.3 Domain-Domain Interactions and Networks

Domain-domain interaction (DDI) networks are useful in protein-domain annotation [5], pathway study [6], and complex disease associations [7]. Hence, comparative analysis [8] and knowledge-guided inference [9] of DDI are relevant in the understanding of protein-protein interaction (PPI) in signal transduction [10] and functional networks [11]. An understanding of DDI using complex PPI [12–13] is both interesting and intriguing in the context of proteome level functional annotation of emerging proteins.

P. Kangueane, C. Nilofer, *Protein-Protein and Domain-Domain Interactions*,
https://doi.org/10.1007/978-981-10-7347-2_12

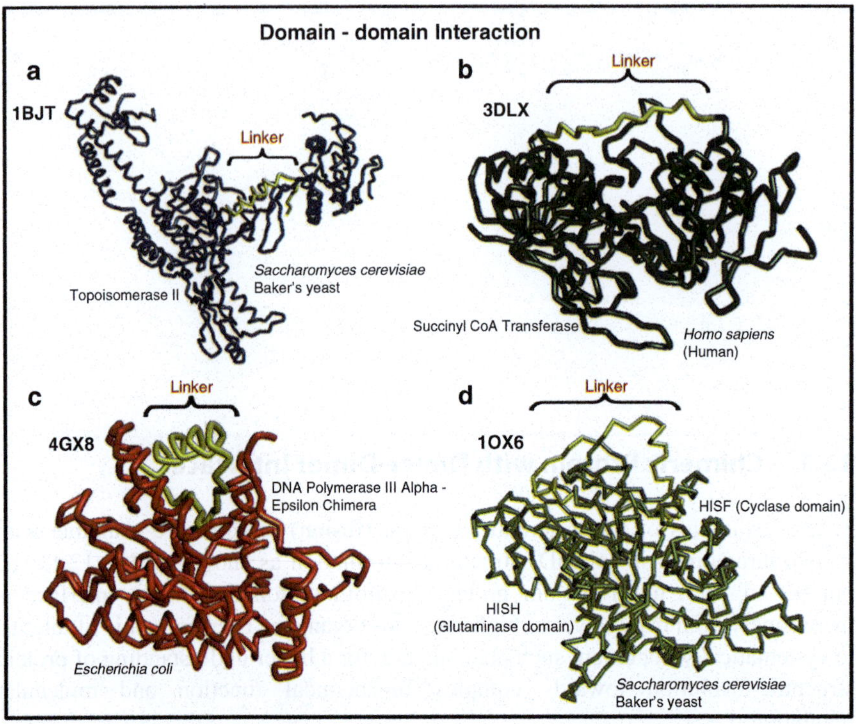

Fig. 12.1 Examples of chimeric proteins with dimer-dimer interface are shown. The domains are connected by linkers as shown in these examples. (**a**) Topoisomerase II; (**b**) succinyl CoA transferase; (**c**) DNA polymerase III epsilon chimeric protein; (**d**) HISH-HISF chimeric protein

12.4 Domain-Domain Interaction Analysis

Domain-domain interaction (DDI) is relevant in functional genomics. Therefore, development of methods for the identification of domain-domain interaction is critical. DDI are determined using segmental labeling of large multi-domain proteins [14]. They are also inferred using PPI analysis [15, 16]. Hence, information on their geometric association [17], conserved orientation [18], spatial orientation [19], and evolutionary conservation [20] is pertinent. Known structural data [21–22] on DDI is highly applicable in the development of prediction models to infer their role in complete genome species organisms [23].

12.5 Prediction of Domain-Domain Interaction

Prediction of DDI is important in genome annotation. A number of methods are available for DDI prediction. Prediction based on parsimony [24, 25], feature discrimination [26] and feature-based support vector machine (SVM) model [27]

leading to the development of an integrated model [28] are available. Methods to predict strong and variants of DDI are also established [29, 30]. The application of DDI in the prediction of PPI networks is significant in functional genomics [31–32].

Exercises

1. Illustrate a domain-domain interface.
2. Show structural examples of domain-domain interface.
3. Give an account of domain-domain interactions in model organism.
4. What are the general features associated with domain-domain interfaces?
5. Name some databases for domain-domain interaction.
6. Name some servers for domain-domain interaction.
7. State the applications of domain-domain interactions.
8. Enumerate the relationship between DDI and PPI.
9. Give an account of DDI in networks.
10. State the association of DDI and diseases.

References

1. Kim Y, Min B, Yi GS. IDDI: integrated domain-domain interaction and protein interaction analysis system. Proteome Sci. 2012;10(Suppl 1): S9.
2. Yellaboina S, Tasneem A, Zaykin DV, Raghavachari B, Jothi R. DOMINE: a comprehensive collection of known and predicted domain-domain interactions. Nucleic Acids Res. 2011;39(Database issue):D730–5.
3. Stein A, Panjkovich A, Aloy P. 3did update: domain-domain and peptide-mediated interactions of known 3D structure. Nucleic Acids Res. 2009;37(Database issue):D300–4.
4. Kundrotas PJ, Alexov E. PROTCOM: searchable database of protein complexes enhanced with domain-domain structures. Nucleic Acids Res. 2007;35(Database issue):D575–9.
5. Zhao XM, Wang Y, Chen L, Aihara K. Protein domain annotation with predicted domain-domain interaction networks. Protein Pept Lett. 2008;15(5):456–62.
6. Ng KL, Huang CH, Liu HC, Liu HC. Applications of domain-domain interactions in pathway study. Comput Biol Chem. 2008;32(2):81–7.
7. Wang W, Zhang W, Jiang R, Luan Y. Prioritisation of associations between protein domains and complex diseases using domain-domain interaction networks. IET Syst Biol. 2010;4(3):212–22.
8. Björkholm P, Sonnhammer EL. Comparative analysis and unification of domain-domain interaction networks. Bioinformatics. 2009;25(22):3020–5.
9. Liu M, Chen XW, Jothi R. Knowledge-guided inference of domain-domain interactions from incomplete protein-protein interaction networks. Bioinformatics. 2009;25(19):2492–9.
10. Nguyen TP, Ho TB. Discovering signal transduction networks using signaling domain-domain interactions. Genome Inform. 2006;17(2):35–45.
11. Schlicker A, Huthmacher C, Ramírez F, Lengauer T, Albrecht M. Functional evaluation of domain-domain interactions and human protein interaction networks. Bioinformatics. 2007;23(7):859–65.

12. Albrecht M, Huthmacher C, Tosatto SC, Lengauer T. Decomposing protein networks into domain-domain interactions. Bioinformatics. 2005;21 Suppl 2: ii220–1.
13. Chen C, Zhao JF, Huang Q, Wang RS, Zhang XS. Inferring domain-domain interactions from protein-protein interactions in the complex network conformation. BMC Syst Biol. 2012;6(Suppl 1): S7.
14. Chen J, Wang JA. Segmental labeling strategy for unambiguous determination of domain-domain interactions of large multi-domain proteins. J Biomol NMR. 2011;50(4):403–10.
15. Deng M, Mehta S, Sun F, Chen T. Inferring domain-domain interactions from protein-protein interactions. Genome Res. 2002;12(10):1540–8.
16. Khor S. Inferring domain-domain interactions from protein-protein interactions with formal concept analysis. PLoS One. 2014;9(2):e88943.
17. Kim WK, Ison JC. Survey of the geometric association of domain-domain interfaces. Proteins. 2005;61(4):1075–88.
18. Littler SJ, Hubbard SJ. Conservation of orientation and sequence in protein domain--domain interactions. J Mol Biol. 2005;345(5):1265–79.
19. Bhaskara RM, de Brevern AG, Srinivasan N. Understanding the role of domain-domain linkers in the spatial orientation of domains in multi-domain proteins. J Biomol Struct Dyn. 2013;31(12):1467–80.
20. Itzhaki Z, Akiva E, Altuvia Y, Margalit H. Evolutionary conservation of domain-domain interactions. Genome Biol. 2006;7(12):R125.
21. Jefferson ER, Walsh TP, Barton GJ. A comparison of SCOP and CATH with respect to domain-domain interactions. Proteins. 2008;70(1):54–62.
22. Zhou H, Xue B, Zhou Y. DDOMAIN: Dividing structures into domains using a normalized domain-domain interaction profile. Protein Sci. 2007;16(5):947–55.
23. Guda C, King BR, Pal LR, Guda P. A top-down approach to infer and compare domain-domain interactions across eight model organisms. PLoS One. 2009;4(3):e5096.
24. Guimarães KS, Jothi R, Zotenko E, Przytycka TM. Predicting domain-domain interactions using a parsimony approach. Genome Biol. 2006;7(11):R104.
25. Guimarães KS, Przytycka TM. Interrogating domain-domain interactions with parsimony based approaches. BMC Bioinform. 2008;9:171.
26. Zhao XM, Chen L, Aihara K. A discriminative approach for identifying domain-domain interactions from protein-protein interactions. Proteins. 2010;78(5):1243–53.
27. González AJ, Liao L. Predicting domain-domain interaction based on domain profiles with feature selection and support vector machines. BMC Bioinformatics. 2010;11:537.
28. Lee H, Deng M, Sun F, Chen T. An integrated approach to the prediction of domain-domain interactions. BMC Bioinformatics. 2006;7:269.
29. Nye TM, Berzuini C, Gilks WR, Babu MM, Teichmann S. Predicting the strongest domain-domain contact in interacting protein pairs. Stat Appl Genet Mol Biol. 2006;5:Article5.
30. Gomes M, Hamer R, Reinert G, Deane CM. Mutual information and variants for protein domain-domain contact prediction. BMC Res Notes. 2012;5:472.
31. Zhao Y, Hayashida M, Nacher JC, Nagamochi H, Akutsu T. Protein complex prediction via improved verification methods using constrained domain-domain matching. Int J Bioinforma Res Appl. 2012;8(3–4):210–27.
32. Ma W, McAnulla C, Wang L. Protein complex prediction based on maximum matching with domain-domain interaction. Biochim Biophys Acta. 2012;1824(12):418–24.

Protein-Protein Interaction Tools

13

13.1 PPI Tools for Protein

Computer-aided mathematical tools (Tables 13.1, 13.2, 13.3, 13.4, 13.5, and 13.6) for the analysis and prediction of PPI are highly relevant in cell and molecular biology experiments. A number of such tools are currently available as shown in Fig. 13.1. These help in the understanding of complex issues related to PPI in biology.

13.2 PPI Prediction Tools

Prediction of interaction partners for proteins is a critical issue in molecular biology. We document a list of such tools in Table 13.1.

13.2.1 BindML+

BindML+ [1] classifies PPI into permanent and transient interfaces using amino acid substitution model.

13.2.2 PIC (Protein Interactions Calculator)

PIC [2] computes intra- and inter-protein interactions like hydrophobic, disulfide bonds, hydrogen bonds, ionic, aromatic-aromatic, aromatic-sulfur, cation-pi, and apolar-apolar residue interactions.

P. Kangueane, C. Nilofer, *Protein-Protein and Domain-Domain Interactions*,
https://doi.org/10.1007/978-981-10-7347-2_13

Table 13.1 PPI prediction tools

S. no.	Tool name	Description	URL	Web/desktop	References
1	BindML+	Predicting permanent and transient interfaces	http://kiharalab.org/bindml/plus/info.html	Web and desktop	[1]
2	PIC	Protein Interactions Calculator	http://pic.mbu.iisc.ernet.in/	Web	[2]
3	MAPPIS	Multiple Alignment of Protein-Protein Interfaces	http://bioinfo3d.cs.tau.ac.il/MAPPIS/	Web	[3]
4	hotPOINT	Hot spot prediction server for protein interfaces	http://prism.ccbb.ku.edu.tr/hotpoint/	Web	[4]
5	PSOPIA	Prediction server for protein-protein interactions	http://mizuguchilab.org/PSOPIA	Web	[5]
6	APID	Agile Protein Interaction DataAnalyzer	http://cicblade.dep.usal.es:8080/APID/init.action	Web	[6]
7	PROTORP	Protein-protein interface analysis server	http://www.cs.cmu.edu/~chongw/citeulike/docs/doc14715.html	Web	[7]
8	iFrag	A protein-protein interface prediction server based on sequence fragments	http://sbi.imim.es/iFrag	Web	[8]
9	Interactome3D	Interactome3D	http://interactome3d.irbbarcelona.org/	Web	[9]
10	PRED_PPI	Predicting protein-protein interactions	http://cic.scu.edu.cn/bioinformatics/predict_ppi/	Web	[10]
11	PCRPi-W	Presaging critical residues in protein interfaces-web server	http://www.bioinsilico.org/PCRPi	Web	[11]
12	PPCheck	Checks the strength of protein-protein interaction	http://caps.ncbs.res.in/ppcheck/	Web	[12]

Table 13.2 PPI binding site prediction tools

S. no.	Tool name	Description	URL	Web/ desktop	References
1	Meta-PPISP	A meta server for protein-protein interaction site prediction	http://pipe.scs.fsu.edu/meta-ppisp.html	Web	[13]
2	BindML	Binding site prediction by maximum likelihood	http://kiharalab.org/bindml/	Web and desktop	[14]
3	Sppider	Solvent accessibility-based protein-protein interface identification and recognition	http://sppider.cchmc.org/	Web	[15]
4	InterProSurf	Interacting sites on protein surfaces	http://curie.utmb.edu/prosurf.html	Web	[16]

Table 13.3 PPI prediction tool for model organisms

S. no.	Tool name	Description	URL	Web/ desktop	References
1	TRI_tool	Human transcriptional regulation PPI prediction	http://www.vin.bg.ac.rs/180/tools/tfpred.php	Web	[17]
2	HIVsemi	Semi-supervised multitask learning for predicting interactions	http://www.cs.cmu.edu/~qyj/HIVsemi	Desktop	[18]
3	ChiPPI	Chimeric protein-protein interaction server	http://chippi.md.biu.ac.il/	Web	[19]
4	HIPPIE	Human Integrated Protein-Protein Interaction Reference	http://cbdm-01.zdv.uni-mainz.de/~mschaefer/hippie/	Web	[20]
5	InteroPORC	Automatic molecular interaction predictions	http://biodev.extra.cea.fr/interoporc/	Web and desktop	[21]

13.2.3 MAPPIS (Multiple Alignment of Protein-Protein Interfaces)

MAPPIS [3] examines protein-protein interfaces for the presence of conserved physicochemical interaction using multiple sequence alignment (MSA).

13.2.4 HotPOINT (Hot Spot Prediction Server for Protein Interfaces)

HotPOINT [4] identifies the hot spot residues (high-energy contributors) at the interfaces of PPI.

Table 13.4 Structure-based PPI prediction tool

S. no.	Tool name	Description	URL	Web/ desktop	References
1	PrISE	Prediction of protein-protein interface residues using structural elements	http://ailab1.ist.psu.edu/prise/index.py	Web	[22]
2	InterPreTS	Interaction prediction through tertiary structure	http://www.russell.embl.de/interprets	Web and desktop	[23]
3	iLoops	Protein-protein interaction prediction server based on structural features	http://sbi.imim.es/iLoopsServer/	Web	[24]
4	Struct2Net	Structure-based computational predictions of protein-protein interactions	http://cb.csail.mit.edu/cb/struct2net/webserver/	Web	[25]
5	PRISM	Protein-protein interactions and modeling their 3D complexes	http://cosbi.ku.edu.tr/prism/	Web	[26]
6	SAPIN	Structural analysis of protein interaction networks	http://sapin.crg.es/	Web	[27]

Table 13.5 PPI tools for network creation

S. no.	Tool name	Description	URL	Web/ desktop	References
1	PPI spider	Interpretation of proteomics data in the context of PPI networks	http://www.bioprofiling.de/PPI_spider.html	Web	[28]
2	Path2PPI	Prediction of pathway-related protein-protein interaction networks	http://bioconductor.org/packages/release/bioc/html/Path2PPI.html	Desktop	[29]
3	POINeT	Protein interactome with subnetwork analysis and hub prioritization	http://poinet.bioinformatics.tw/	Web	[30]
4	MyProteinNet	Create context-sensitive protein interaction networks	http://netbio.bgu.ac.il/myproteinnet	Web	[31]
5	RedNemo	Repeated diffusion with neighborhood modifications	http://webprs.khas.edu.tr/~cesim/RedNemo.tar.gz	Desktop	[32]

Table 13.6 Other PPI-related tools

S. no.	Tool name	Description	URL	Web/desktop	References
1	FunFOLD2	Server for the prediction of protein-ligand interactions	http://www.reading.ac.uk/bioinf/FunFOLD/	Desktop	[33]
2	LigPlot+	Multiple ligand-protein interaction diagrams for drug discovery	http://www.ebi.ac.uk/thornton-srv/software/LIGPLOT/	Desktop	[34]
3	LIGPLOT	Program for automatically plotting protein-ligand interactions	http://www.ebi.ac.uk/thornton-srv/software/LIGPLOT/	Desktop	[35]
4	PepSite	Prediction on peptide-binding sites and protein surfaces	http://pepsite2.russelllab.org/	Web	[36]
5	PPIMpred	High-throughput screening of small molecules targeting PPI	http://bicresources.jcbose.ac.in/ssaha4/PPIMpred/	Web	[37]

13.2.5 PSOPIA (Prediction Server of Protein-Protein Interactions)

PSOPIA [5] is a computational method to calculate PPI for given protein sequence pair using averaged one-dependence estimators (AODE) along with the features of homologous PPIs.

13.2.6 APID (Agile Protein Interaction DataAnalyzer)

APID [6] is a collection of tools to exhibit all related information for the given interaction details. It demonstrates molecular interaction details of PPI using information from six main databases (BIND, BioGRID, DIP, HPRD, IntAct, and MINT as discussed in Chap. 9).

13.2.7 PROTORP (Protein-Protein Interaction Analysis Server)

PROTORP [7] analyzes PPI using protein structure to decipher its physical and chemical properties.

13.2.8 iFrag

iFrag [8] predicts binding between two protein sequences using scoring matrix for similar residue fragments existing at the interface region of the proteins.

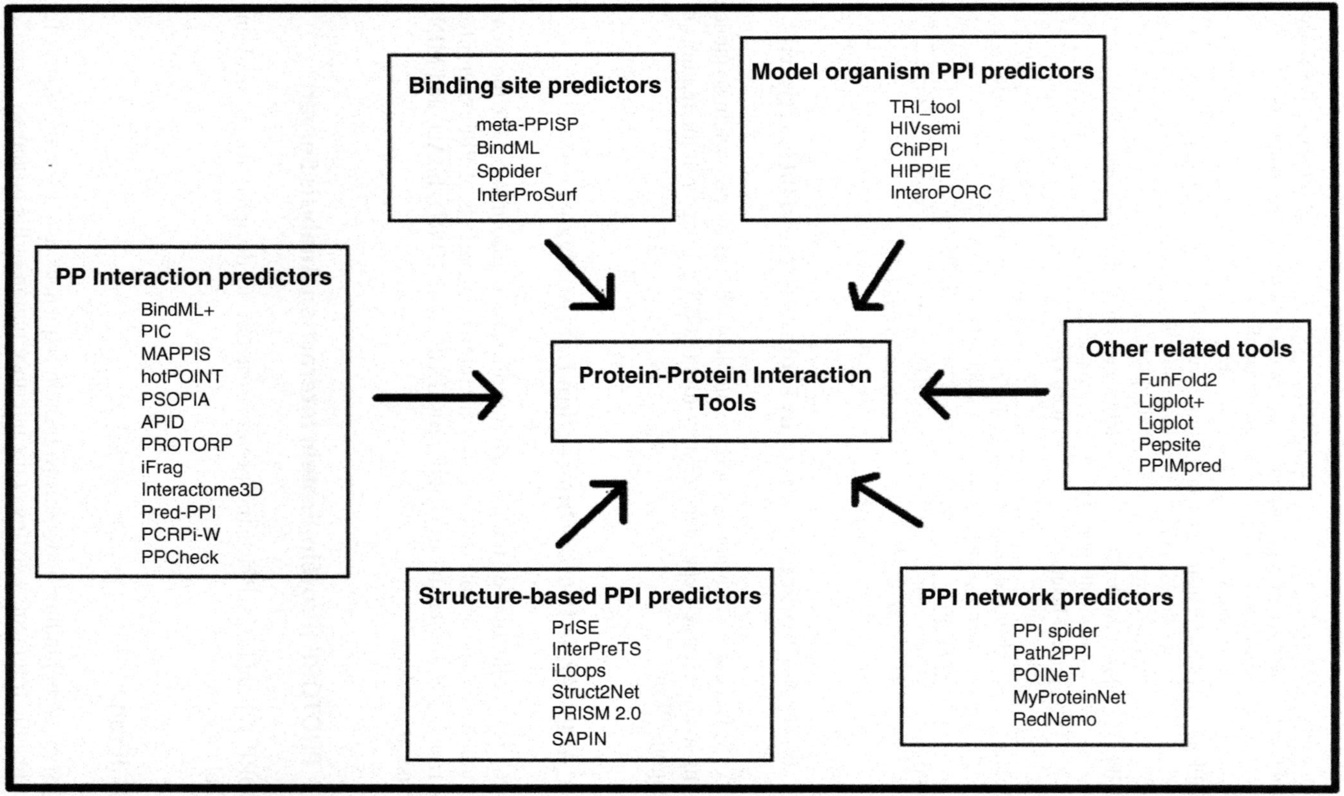

Fig. 13.1 Tools for PPI analysis and prediction grouped into different categories. These include (1) protein partner prediction, PPI site prediction, structure-driven PPI prediction, PPI network prediction, and proteome-scale analysis in model organisms

13.2.9 Interactome3D

Interactome3D [9] annotates and models PPI using information from major pathway databases.

13.2.10 PRED_PPI (Predicting Protein-Protein Interaction)

PRED_PPI [10] computes PPI for five organisms, namely, humans, yeast, *Drosophila*, *Escherichia coli*, and *Caenorhabditis elegans* using support vector machine (SVM) with a probability of 0.5 as cutoff.

13.2.11 PCRPi-W (Presaging Critical Residues in Protein Interfaces-Web Server)

PCRPi-W [11] uses Bayesian networks to predict critical high binding energy residues (hot spots) at the interface.

13.2.12 PPCheck

PPCheck [12] measures strength of PPI quantitatively using pseudo-energy assignment, and it also predicts hot spot residues at the interface region.

13.3 Tool for Predicting PPI Binding Sites

Tools for predicting specific binding sites between proteins are useful to decipher their molecular function as given in Table 13.2.

13.3.1 Meta-PPISP

Meta-PPISP [13] is a tool with high accuracy to find PPI binding sites by combining results of various predictors and validating the results using linear regression method.

13.3.2 BindML

BindML [14] employs phylogenetic substitution method to identify the residues in the binding interface of PPI.

13.3.3 Sppider (Solvent Accessibility-Based Protein-Protein Interface Identification and Recognition)

Sppider [15] predicts protein-protein interaction sites using relative solvent accessibility (RSA) of the amino acid residues to identify binding sites.

13.3.4 InterProSurf

InterProSurf [16] is a computational method used to predict residues in the interface region that overlap over a number of interaction with different proteins. Interface binding residues are determined using sequence analysis and mutagenesis experiment.

13.4 Tools for PPI Analysis in Model Organism

Proteome-level understanding of PPI is highly useful in deciphering genome evolution. We document a list of such tools in Table 13.3.

13.4.1 TRI_tool (Transcriptional Regulation Interactions)

TRI_tool [17] detects PPI among human transcription regulators.

13.4.2 HIVsemi

HIVsemi [18] identifies PPI binding residues between HIV-1 and human using semi-supervised multitask learning.

13.4.3 ChiPPI (Chimeric Protein-Protein Interaction)

ChiPPI [19] is a comprehensive tool to discover PPI network existing in the fusion proteins of cancer by mapping chimeric protein-protein interactions.

13.4.4 HIPPIE (Human Integrated Protein-Protein Interaction Reference)

HIPPIE [20] generates query-specific human PPI networks with interpretation from literature.

13.4.5 InteroPORC (Interaction and Clusters of Orthologous Proteins)

InteroPORC [21] predicts PPI networks in model organisms.

13.5 Tools for Structure-Based PPI Prediction

Known protein complexes provide useful information to help in predicting PPI. We document a list of such tools in Table 13.4.

13.5.1 PrISE (Prediction of Protein-Protein Interface Residues Using Structural Elements)

PrISE [22] uses structural elements to identify binding residues in the interface region of PPI.

13.5.2 InterPreTS (Interaction Prediction Through Tertiary Structure)

InterPreTS [23] predicts interaction by discovering the given sequence using database of interacting domains (DBID) and checks for conserved residues at the interface.

13.5.3 iLoops

iLoops [24] predicts PPI-based structural features using random forest classifiers.

13.5.4 Struct2Net

Struct2Net [25] is a structure-based PPI prediction tool.

13.5.5 PRISM 2.0 (Protein Interactions by Structural Matching 2.0)

PRISM [26] is a structural model and knowledge-based tool to identify PPI and its interface residues.

13.5.6 SAPIN (Structural Analysis of Protein Interaction Networks)

SAPIN [27] is a framework to perform structural analysis on PPI network to identify the relationship between proteins.

13.6 Tools for Predicting PPI Networks

Protein-protein interaction results in networks. Therefore, it is of interest to develop tools for predicting PPI networks as given in Table 13.5.

13.6.1 PPI Spider

PPI spider [28] exploits experimentally driven protein data to predict PPI and uses Monte Carlo simulation to infer it by statistical significance.

13.6.2 Path2PPI

Path2PPI [29] is an R package to identify PPI networks for organisms with full genome sequence.

13.6.3 POINeT

POINeT [30] finds PPI's network and subnetwork with data from multiple resources using ranking filter method (S3). This finding helps in identifying novel markers for prevailing diseases.

13.6.4 MyProteinNet

MyProteinNet [31] allows user to build a query-specific protein network by gathering data from 11 different databases to create protein-protein interaction network.

13.6.5 RedNemo (Repeated Diffusion with Neighborhood Modifications)

RedNemo [32] develops accurate topology-based PPI network reconstruction to overcome the existing scalability issue.

13.7 Other PPI-Related Tools

It should be noted that there are other PPI-related tools as given in Table 13.6. These tools are indirectly associated with the study of PPI.

13.7.1 FunFold2

FunFold2 [33] infers protein-ligand interaction and predicts protein function with query sequence having quality assessment.

13.7.2 Ligplot+

Ligplot+ [34] generates 2D diagrams for multiple protein-ligand interaction using given query structure to predict interaction patterns and relationship.

13.7.3 Ligplot

Ligplot [35] generates 2D diagrams for protein-ligand interaction to infer different kinds of existing interaction.

13.7.4 Pepsite

Pepsite [36] identifies peptide-binding spots in the protein-peptide interaction using the query sequence data at a fast execution speed.

13.7.5 PPIMpred (Prediction of Protein-Protein Interaction Modulators)

PPIMpred [37] predicts small molecules (ligands) through high-throughput screening to target PPI and inhibit its action using SVM method.

13.8 Conclusion

The available tools have helped in the comprehensive understanding of PPI under different circumstances. We foresee more such tools in the future for establishing more clarity on this issue.

Exercises

1. Give an account of PPI tools.
2. State the different types of PPI tools.
3. What is an online web server?
4. What is stand-alone software?
5. What is a computer-aided tool?
6. What is a protein-protein interface?
7. Give an account of binding site prediction in protein complexes.
8. What are protein-protein interaction networks?
9. Give an account of proteome-scale analysis of PPI in model organism.
10. Name some structure-based PPI prediction tools.

References

1. La D, Kong M, Hoffman W, Choi YI, Kihara D. Predicting permanent and transient protein-protein interfaces. Proteins. 2013;81(5):805–18.
2. Tina KG, Bhadra R, Srinivasan N. PIC: Protein interactions calculator. Nucleic Acids Res. 2007;35(Web Server issue):W473–6.
3. Shulman-Peleg A, Shatsky M, Nussinov R, Wolfson HJ. MultiBind and MAPPIS: webservers for multiple alignment of protein 3D-binding sites and their interactions. Nucleic Acids Res. 2008;36(Web Server issue):W260–4.
4. Tuncbag N, Keskin O, Gursoy A. HotPoint: hot spot prediction server for protein interfaces. Nucleic Acids Res. 2010;38(Web Server issue):W402–6.
5. Murakami Y, Mizuguchi K. Homology-based prediction of interactions between proteins using averaged one-dependence estimators. BMC Bioinf. 2014;15:213.
6. Prieto C, De Las Rivas J. APID: Agile protein interaction data analyzer. Nucleic Acids Res. 2006;34(Web Server issue):W298–302.
7. Reynolds C, Damerell D, Jones S. ProtorP: a protein-protein interaction analysis server. Bioinformatics. 2009;25(3):413–4.
8. Garcia-Garcia J, Valls-Comamala V, Guney E, Andreu D, Muñoz FJ, Fernandez-Fuentes N, et al. iFrag: a protein-protein Interface prediction server based on sequence fragments. J Mol Biol. 2017;429(3):382–9.
9. Mosca R, Céol A, Aloy P. Interactome3D: adding structural details to protein networks. Nat Methods. 2013;10(1):47–53.
10. Guo Y, Li M, Pu X, Li G, Guang X, Xiong W, et al. PRED_PPI: a server for predicting protein-protein interactions based on sequence data with probability assignment. BMC Res Notes. 2010;3:145.
11. Segura Mora J, Assi SA, Fernandez-Fuentes N. Presaging critical residues in protein interfaces-web server (PCRPi-W): a web server to chart hot spots in protein interfaces. PLoS One. 2010;5(8):e12352.
12. Sukhwal A, Sowdhamini R. PPCheck: a webserver for the quantitative analysis of protein-protein interfaces and prediction of residue hotspots. Bioinform Biol Insights. 2015;9:141–51.
13. Qin S, Zhou HX. Meta-PPISP: a meta web server for protein-protein interaction site prediction. Bioinformatics. 2007;23(24):3386–7.
14. La D, Kihara DA. Novel method for protein-protein interaction site prediction using phylogenetic substitution models. Proteins. 2012;80(1):126–41.
15. Porollo A, Meller J. Prediction-based fingerprints of protein-protein interactions. Proteins. 2007;66(3):630–45.

16. Negi SS, Schein CH, Oezguen N, Power TD, Braun W. InterProSurf: a web server for predicting interacting sites on protein surfaces. Bioinformatics. 2007;23(24):3397–9.
17. Perovic V, Sumonja N, Gemovic B, Toska E, Roberts SG, Veljkovic N. TRI_tool: a web-tool for prediction of protein-protein interactions in human transcriptional regulation. Bioinformatics 2017 33(2): 289–291.
18. Qi Y, Tastan O, Carbonell JG, Klein-Seetharaman J, Weston J. Semi-supervised multi-task learning for predicting interactions between HIV-1 and human proteins. Bioinformatics. 2010;26(18):i645–52.
19. Frenkel-Morgenstern M, Gorohovski A, Tagore S, Sekar V, Vazquez M, Valencia A. ChiPPI: a novel method for mapping chimeric protein-protein interactions uncovers selection principles of protein fusion events in cancer. Nucleic Acids Res. 2017;
20. Alanis-Lobato G, Andrade-Navarro MA, Schaefer MH. HIPPIE v2.0: enhancing meaningfulness and reliability of protein-protein interaction networks. Nucleic Acids Res. 2017;45(D1):D408–14.
21. Michaut M, Kerrien S, Montecchi-Palazzi L, Chauvat F, Cassier-Chauvat C, Aude JC, et al. InteroPORC: automated inference of highly conserved protein interaction networks. Bioinformatics. 2008;24(14):1625–31.
22. Jordan RA, El-Manzalawy Y, Dobbs D, Honavar V. Predicting protein-protein interface residues using local surface structural similarity. BMC Bioinform. 2012;13:41.
23. Aloy P, Russell RB. InterPreTS: protein interaction prediction through tertiary structure. Bioinformatics. 2003;19(1):161–2.
24. Planas-Iglesias J, Marin-Lopez MA, Bonet J, Garcia-Garcia J, Oliva B. iLoops: a protein-protein interaction prediction server based on structural features. Bioinformatics. 2013;29(18):2360–2.
25. Singh R, Park D, Xu J, Hosur R, Berger B. Struct2Net: a web service to predict protein-protein interactions using a structure-based approach. Nucleic Acids Res. 2010;38(Web Server issue):W508–15.
26. Baspinar A, Cukuroglu E, Nussinov R, Keskin O, Gursoy A. PRISM: a web server and repository for prediction of protein-protein interactions and modeling their 3D complexes. Nucleic Acids Res. 2014;42(Web Server issue):W285–9.
27. Yang JS, Campagna A, Delgado J, Vanhee P, Serrano L, Kiel C. SAPIN: a framework for the structural analysis of protein interaction networks. Bioinformatics. 2012;28(22):2998–9.
28. Antonov AV, Dietmann S, Rodchenkov I, Mewes HW. PPI spider: a tool for the interpretation of proteomics data in the context of protein-protein interaction networks. Proteomics. 2009;9(10):2740–9.
29. Philipp O, Osiewacz HD, Koch I. Path2PPI: an R package to predict protein-protein interaction networks for a set of proteins. Bioinformatics. 2016;32(9):1427–9.
30. Lee SA, Chan CH, Chen TC, Yang CY, Huang KC, Tsai CH. POINeT: protein interactome with sub-network analysis and hub prioritization. BMC Bioinform. 2009;10:114.
31. Basha O, Flom D, Barshir R, Smoly I, Tirman S, Yeger-Lotem E. MyProteinNet: build up-to-date protein interaction networks for organisms, tissues and user-defined contexts. Nucleic Acids Res. 2015;43(W1):W258–63.
32. Alkan F, Erten C. RedNemo: topology-based PPI network reconstruction via repeated diffusion with neighborhood modifications. Bioinformatics. 2017;33(4):537–44.
33. Roche DB, Buenavista MT, McGuffin LJ. The FunFOLD2 server for the prediction of protein-ligand interactions. Nucleic Acids Res. 2013;41(Web Server issue):W303–7.
34. Laskowski RA, Swindells MB. LigPlot+: multiple ligand-protein interaction diagrams for drug discovery. J Chem Inf Model. 2011;51(10):2778–86.
35. Wallace AC, Laskowski RA, Thornton JM. LIGPLOT: a program to generate schematic diagrams of protein-ligand interactions. Protein Eng. 1995;8(2):127–34.
36. Trabuco LG, Lise S, Petsalaki E, Russell RB. PepSite: prediction of peptide-binding sites from protein surfaces. Nucleic Acids Res. 2012;40(Web Server issue):W423–7.
37. Jana T, Ghosh A, Das Mandal S, Banerjee R, Saha S. PPIMpred: a web server for high-throughput screening of small molecules targeting protein-protein interaction. R Soc Open Sci. 2017;4(4):160501.

Protein-Protein Docking: Methods and Tools

14

14.1 Protein-Protein Docking

Protein-protein docking is the prediction of binding between two protein structures so as to form a protein complex using features such as steric and physicochemical complementarity at the protein-protein interface. This involves the prediction of conformational changes between unbound and bound structures. There are a number of scientific groups involved in the development and assessments of predictive methodologies for protein-protein binding. This is possible using structural data with a deeper understanding of the fundamental principles of protein interactions together with available advanced computational capabilities.

14.2 CAPRI

CAPRI refers to Critical Assessment of Predicted Interactions. CAPRI is a scientific meeting organized every 6 months to discuss protein-protein docking accuracy by a double-blind method planned between theoretical docking groups and X-ray crystallographers of protein complexes where the targets are predefined, agreed upon, and compared. A list of available docking tools is illustrated in Fig. 14.1.

14.3 CAPRI Validated Models and Tools

The list of CAPRI validated models and tools is given in Table 14.1.

P. Kangueane, C. Nilofer, *Protein-Protein and Domain-Domain Interactions*,
https://doi.org/10.1007/978-981-10-7347-2_14

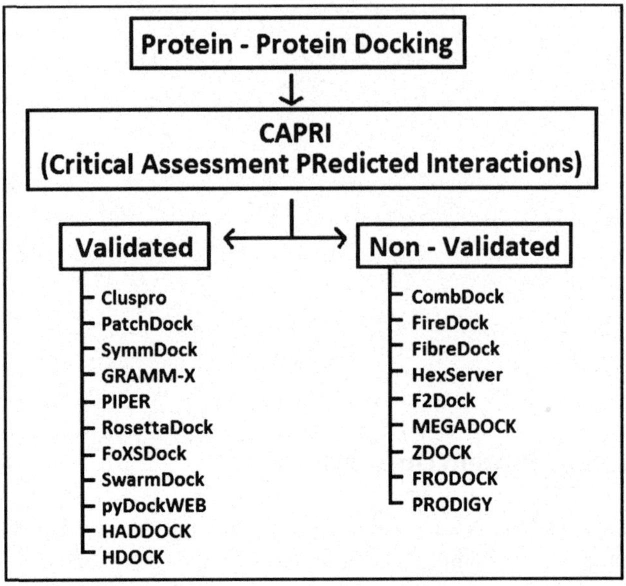

Fig. 14.1 A list of CAPRI validated and non-validated tools is shown in this graphical illustration

14.3.1 ClusPro

ClusPro [1] is an automated web-based docking software that uses clustering method to rank the docked protein complexes with better electrostatic and de-solvation free energy.

14.3.2 PatchDock and SymmDock

PatchDock and SymmDock [2] are two docking web servers that predict the structure of protein complexes using shape complementarity principles and geometry-based docking optimization, respectively.

14.3.3 GRAMM-X

GRAMM-X [3] is a docking web interface based on fast Fourier transform (FFT). FFT includes smoothed potentials, refinement, and knowledge-based scoring.

14.3.4 PIPER

PIPER [4] is a web-based FFT docking program with structure-based pair-wise intermolecular potentials called DARS (Decoys As the Reference State).

Table 14.1 CAPRI validated protein-protein docking tools

S. no.	Docking tool	Description	Method/approach	URL	CAPRI validated	Web/desktop	Year	References
1	ClusPro	Web-based program for computational docking	Fast Fourier transform (FFT) correlation techniques	https://cluspro.bu.edu/login.php	Yes	Web	2004	[1]
2	PatchDock	Structure prediction of PPI complexes	Shape Complementarity Principles and symmetry restrictions	http://bioinfo3d.cs.tau.ac.il/PatchDock/	Yes	Web and desktop	2005	[2]
3	SymmDock	Predicts the structure of a homo multimer with cyclic symmetry		http://bioinfo3d.cs.tau.ac.il/SymmDock/	Yes	Web and desktop	2005	[2]
4	GRAMM-X	GRAMM-X PP Docking Web Server v.1.2.0	Fast Fourier transform (FFT) GRAMM methodology	http://vakser.compbio.ku.edu/resources/gramm/grammx/	Yes	Web	2006	[3]
5	PIPER	Protein-protein docking program	FFT-based with pair-wise potentials	https://www.schrodinger.com/piper	Yes	Desktop	2006	[4]
6	RosettaDock	Local protein-protein docking	Monte Carlo-based algorithm	http://rosettadock.graylab.jhu.edu	Yes	Web	2008	[5]
7	FoXSDock	Macromolecular docking with SAXS profile	Rigid global modeling and flexible docking	http://modbase.compbio.ucsf.edu/foxsdock/	Yes	Web and desktop	2011	[6]
8	SwarmDock	Flexible protein-protein docking	flexible protein-protein docking algorithm	http://bmm.cancerresearchuk.org/%7ESwarmDock/	Yes	Web	2013	[7]
9	pyDockWEB	Structural prediction of PPI	Rigid-body docking and scoring by pyDock	https://life.bsc.es/servlet/pydock/home/	Yes	Web	2013	[8]
10	HADDOCK	High Ambiguity Driven biomolecular DOCKing	Flexible docking	http://haddock.chem.uu.nl/services/HADDOCK	Yes	Web	2016	[9]
11	HDOCK	Web server for PPI	Hybrid docking algorithm of template-based	http://hdock.phys.hust.edu.cn/	Yes	Web	2017	[10]

CAPRI Critical Assessment of Predicted Interactions, *PPI* protein-protein interaction

14.3.5 RosettaDock

RosettaDock [5] is a web-based docking server that employs optimization of rigid-body orientation and side chain conformation to rank the docked protein molecules.

14.3.6 FoXSDock

FoXSDock [6] is a web-based docking method that uses rigid-body docking and SAXS (small angle X-ray scattering) profile to obtain better docked protein complexes with lowest energy.

14.3.7 SwarmDock

SwarmDock [7] is a flexible docking algorithm-based web server that produces docked complexes with low-energy conformation.

14.3.8 pyDockWEB

pyDockWEB [8] is a web server based on rigid-body docking, and the docked complexes are scored using electrostatic and de-solvation energy.

14.3.9 HADDOCK

HADDOCK [9] is a web server based on information-driven docking approach to predict the structures of the protein complex.

14.3.10 HDOCK

HDOCK [10] is a hybrid docking algorithm based web server that performs both modeling and docking with given PDB structures to produce good docked complexes.

14.4 CAPRI Non-validated Models and Tools

The list of CAPRI non-validated models and tools is given in Table 14.2.

Table 14.2 CAPRI non-validated protein-protein docking tools

S. no.	Docking tool	Description	Method/approach	URL	CAPRI validated	Web/ desktop	Year	References
1	CombDock	Combinatorial assembly by multiple docking	Combinatorial algorithm approach	http://bioinfo3d.cs.tau.ac.il/CombDock/download/	No	Web and desktop	2005	[11]
2	FireDock	Fast interaction refinement	Refinement and rescoring of rigid-body PPI	http://bioinfo3d.cs.tau.ac.il/FireDock/	No	Web	2008	[12]
3	FibreDock	Flexible induced-fit backbone refinement	Flexible refinement and rescoring of rigid-body	http://bioinfo3d.cs.tau.ac.il/FiberDock/	No	Web	2010	[13]
4	HexServer	Docking server powered by graphics processors	Fast Fourier transform (FFT) method	http://hexserver.loria.fr/	No	Web	2010	[14]
5	F2Dock	Fast Fourier docking	Fast Fourier transform-based docking algorithm	http://www.cs.utexas.edu/~bajaj/cvc/software/f2dock.shtml	No	Desktop	2011	[15]
6	MEGADOCK	Ultrahigh performance using heterogeneous supercomputers	FFT-grid-based protein-protein docking	http://www.bi.cs.titech.ac.jp/megadock/	No	Desktop	2014	[16]
7	ZDOCK	Prediction of PP complexes and symmetric multimers	Fast Fourier transform (FFT) method	http://zdock.umassmed.edu	No	Web	2014	[17]
8	FRODOCK	Fast Rotational Docking	Complementary knowledge-based potential	http://frodock.chaconlab.org/	No	Desktop	2016	[18]
9	PRODIGY	Protein binding energy	Structural determinants of PPI	http://milou.science.uu.nl/services/PRODIGY	No	Web	2016	[19]

PP protein-protein

14.4.1 CombDock (Combinatorial Docking)

CombDock [11] uses combinatorial algorithm approach to perform protein-protein docking. The contribution made by many associates of protein-protein pair-wise interactions is assembled and predicted.

14.4.2 FireDock (Fast Interaction Refinement)

FireDock [12] is the first flexible docking web server for refining and scoring of docked protein complexes. It optimizes side chain conformation and rigid-body orientation.

14.4.3 FibreDock (Flexible Induced-Fit Backbone Refinement)

FibreDock [13] is the first docking web server used to refine the docked molecules by considering the flexibility of both side chains and backbone.

14.4.4 HexServer

HexServer [14] is a docking server based on fast Fourier transform (FFT) using two graphics processors to speed up the docking process. It is achieved by gaining binding site information of both proteins to speed up the calculation while docking two protein complexes.

14.4.5 F2Dock (Fast Fourier Docking)

F2Dock [15] is a fast Fourier transform-based protein-protein docking tool. It uses a multi-threading method to speed up the execution incorporated with Lennard-Jones potential to rescore the docked protein complexes based on de-solvation energy.

14.4.6 MEGADOCK

MEGADOCK [16] is a FFT-based ultrahigh performance protein-protein docking software. It uses heterogeneous supercomputers to speed up the docking process.

14.4.7 ZDOCK

ZDOCK [17] is a user-friendly and rigid docking-based protein docking server that predicts the structures of protein complexes and symmetric multimers.

14.4.8 FRODOCK (Fast Rotational Docking)

FRODOCK [18] is a docking server that uses complementary knowledge-based potential as a strategy to dock two proteins.

14.4.9 PRODIGY (Protein Binding Energy Prediction)

PRODIGY [19] is a web server used to find the binding affinity between the docked proteins using their intermolecular contacts and properties derived from non-interface surface.

Exercises

1. Expand CAPRI.
2. Describe CAPRI and its purpose.
3. State some CAPRI validated docking tools.
4. State some CAPRI non-validated docking tools.
5. Discuss protein-protein docking techniques.
6. What are the basis of protein-protein docking?
7. Explain flexible docking.
8. Explain rigid docking.
9. Expand FFT and discuss in detail.

References

1. Comeau SR, Gatchell DW, Vajda S, Camacho CJ. ClusPro: a fully automated algorithm for protein-protein docking. Nucleic Acids Res. 2004;32(Web Server issue): W96–9.
2. Schneidman-Duhovny D, Inbar Y, Nussinov R, Wolfson HJ. PatchDock and SymmDock: servers for rigid and symmetric docking. Nucleic Acids Res. 2005;33(Web Server issue):W363–7.
3. Tovchigrechko A, Vakser IA. GRAMM-X public web server for protein-protein docking. Nucleic Acids Res. 2006;34(Web Server issue):W310–4.
4. Kozakov D, Brenke R, Comeau SR, Vajda S. PIPER: an FFT-based protein docking program with pairwise potentials. Proteins. 2006;65(2):392–406.
5. Lyskov S, Gray JJ. The RosettaDock server for local protein-protein docking. Nucleic Acids Res. 2008;36(Web Server issue):W233–8.
6. Schneidman-Duhovny D, Hammel M, Sali A. Macro molecular docking restrained by a small angle X-ray scattering profile. J Struct Biol. 2011;173(3):461–71.
7. Torchala M, Moal IH, Chaleil RA, Fernandez-Recio J, Bates PA. SwarmDock: a server for flexible protein-protein docking. Bioinformatics. 2013;29(6):807–9.
8. Jimenez-Garcia B, Pons C, Fernandez-Recio J. pyDockWEB: a web server for rigid-body protein-protein docking using electrostatics and desolvation scoring. Bioinformatics. 2013;29(13):1698–9.

9. van Zundert GCP, Rodrigues JPGLM, Trellet M, Schmitz C, Kastritis PL, Karaca E, et al. The HADDOCK2.2 web server: user-friendly integrative modeling of biomolecular complexes. J Mol Biol. 2016;428(4):720–5.
10. Yan Y, Zhang D, Zhou P, Li B, Huang SY. HDOCK: web server for protein-protein and protein-DNA/RNA docking based on hybrid strategy. Nucleic Acids Res. 2017;
11. Inbar Y, Benyamini H, Nussinov R, Wolfson HJ. Combinatorial docking approach for structure prediction of large proteins and multi-molecular assemblies. Phys Biol. 2005;2(4):S156–65.
12. Mashiach E, Schneidman-Duhovny D, Andrusier N, Nussinov R, Wolfson HJ. FireDock: a web server for fast interaction refinement in molecular docking. Nucleic Acids Res. 2008;36(Web Server issue):W229–32.
13. Mashiach E, Nussinov R, Wolfson HJ. FiberDock: a web server for flexible induced-fit backbone refinement in molecular docking. Nucleic Acids Res. 2010;38(Web Server issue):W457–61.
14. Macindoe G, Mavridis L, Venkatraman V, Devignes MD, Ritchie DW. HexServer: an FFT-based protein docking server powered by graphics processors. Nucleic Acids Res. 2010;38(Web Server issue):W445–9.
15. Bajaj C, Chowdhury R, Siddavanahalli V. F2Dock: fast Fourier protein-protein docking. IEEE/ACM Trans Comput Biol Bioinform. 2011;8(1):45–58.
16. Ohue M, Shimoda T, Suzuki S, Matsuzaki Y, Ishida T, Akiyama Y. MEGADOCK 4.0: ultra-high-performance protein-protein docking software for heterogeneous supercomputers. Bioinformatics. 2014;30(22):3281–3.
17. Pierce BG, Wiehe K, Hwang H, Kim BH, Vreven T, Weng Z. ZDOCK server: interactive docking prediction of protein-protein complexes and symmetric multimers. Bioinformatics. 2014;30(12):1771–3.
18. Ramírez-Aportela E, López-Blanco JR, Chacón P. FRODOCK 2.0: fast protein-protein docking server. Bioinformatics. 2016;32(15):2386–8.
19. Xue LC, Rodrigues JP, Kastritis PL, Bonvin AM, Vangone A. PRODIGY a web server for predicting the binding affinity of protein-protein complexes. Bioinformatics. 2016;32(23):3676–8.

HLA-Peptide Interaction to Short Peptide Vaccine Design

15

15.1 Short Peptide Vaccine Design

Short peptide vaccine works on the principle of short peptide antigen peptide binding to specific HLA molecules to elicit T-cell immune response through HLA-peptide-TCR complex formation as shown in Fig. 15.1. The short antigen peptide specifically binds to the groove in HLA molecules as shown in Fig. 15.2. The binding of peptides to the HLA molecules at the A to F pockets is shown in Fig. 15.3. There is an induced fit at the binding groove with the peptide as shown in Fig. 15.4.

15.2 HLA Polymorphism

HLA molecules are highly polymorphic. This affects the specific binding of peptides with HLA alleles. There are more than 16 thousand alleles defined so far as shown in Fig. 15.5. This consists of 12 thousand class I alleles and 4 thousand class II alleles as available at the IMGT/HLA database. This is distributed among the sampled ethnic groups of American Indian, Australian Aboriginal, Black, Caucasoid, Hispanic, Mixed, Oriental, and Pacific Islander in the global population of 7.5 billion people. The polymorphism among the known alleles HLA A*0201, A*6801, B*0801, B*2705, B*2709, B*3501, B*5101, B*5301, CW*0304, CW*0401, and E*0101 is shown using a multiple sequence alignment (MSA) diagram in Fig. 15.6. This MSA representation shows sequence-level polymorphism among known HLA alleles along sequence residue position.

P. Kangueane, C. Nilofer, *Protein-Protein and Domain-Domain Interactions*,
https://doi.org/10.1007/978-981-10-7347-2_15

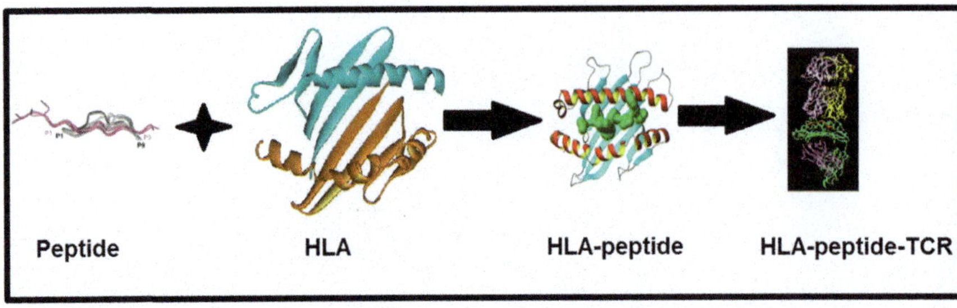

Fig. 15.1 An illustration of HLA-peptide followed by HLA-peptide-TCR binding as a rate-limiting step in T-cell-mediated immune response is shown

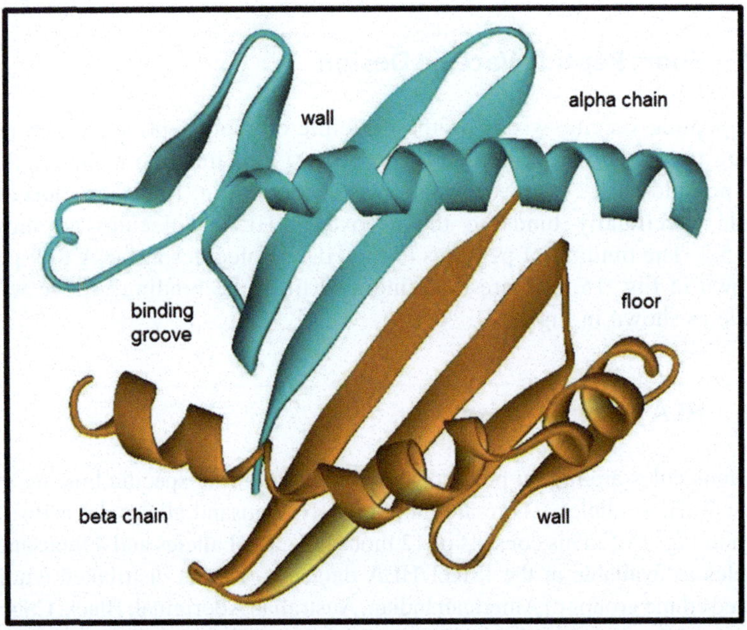

Fig. 15.2 The peptide-binding groove of class I HLA allele formed by domain 1 and domain 2 of the alpha chain is shown. The walls are made of alpha helices, and the floor is made of beta sheets

15.3 Functional HLA Informatics

There are more than 16,000 HLA alleles defined so far at the IMGT/HLA database. These alleles do not bind to a given peptide with identical affinity. It is difficult to quantify the functional variation between them even though they are functionally

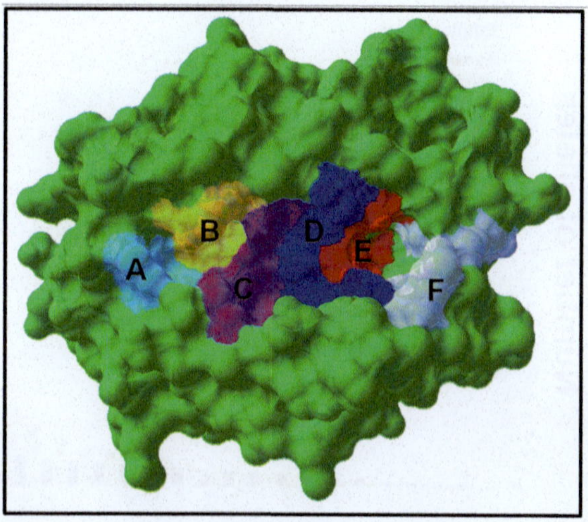

Fig. 15.3 The peptide-binding pockets (*A* to *F*) in class I HLA molecule are shown

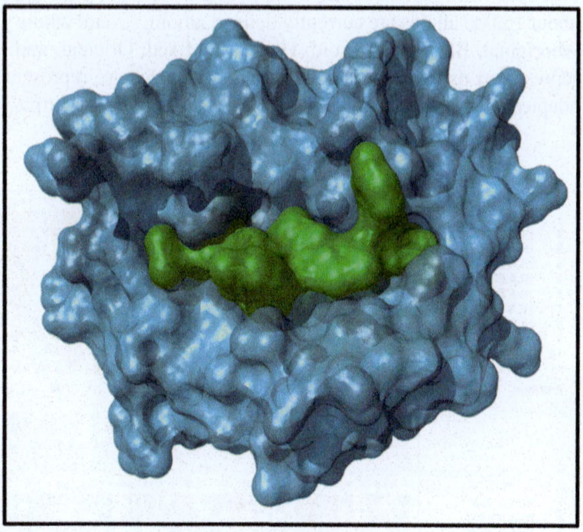

Fig. 15.4 The peptide binding with the binding groove of class I HLA molecule is shown using molecular surface representation. It should be noted that the peptide fits very well within the binding pockets of HLA

related. Therefore, it is important to document their functional similarity by realizing the compression of sequence diversity among HLA alleles [1]. This is further demonstrated by clustering class I HLA alleles using electrostatic distribution maps of the peptide-binding grooves [2] as shown in Fig. 15.7. This grouping provides valuable insight to infer functional overlap among HLA alleles.

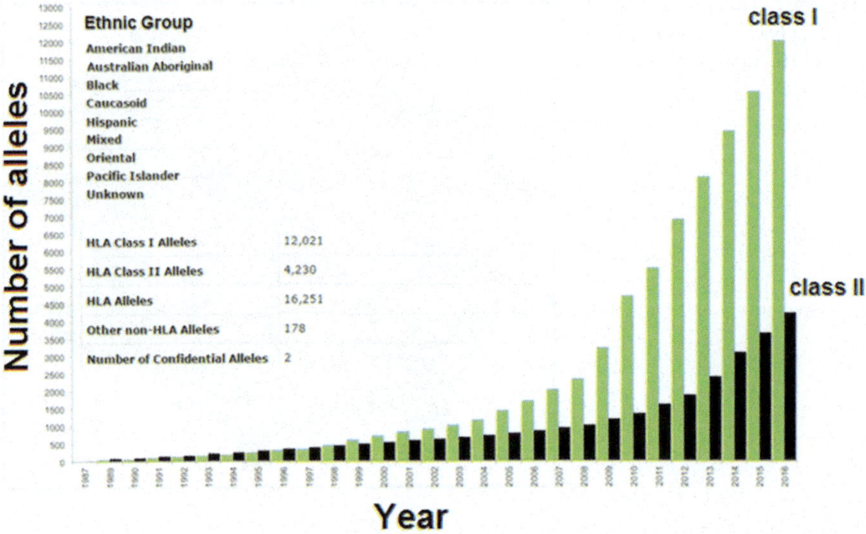

Fig. 15.5 The growth of HLA alleles over a period of three decades is shown. It should be noted that about 16,000 alleles are currently defined among several ethnic groups (American Indian, Australian Aboriginal, Black, Caucasoid, Hispanic, Mixed, Oriental, and Pacific Islander). Class I alleles are represented using green bars, and class II alleles are represented using black bars (This image is adapted from IPD-IMGT/HLA available at URL address http://www.ebi.ac.uk/ipd/imgt/hla/)

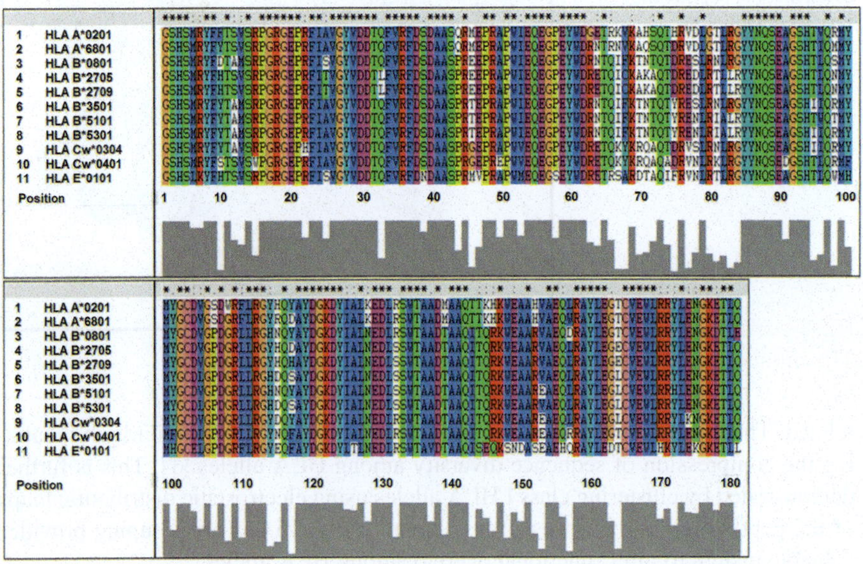

Fig. 15.6 Multiple sequence alignment (MSA) showing residue-level polymorphism among HLA A*0201, A*6801, B*0801, B*2705, B*2709, B*3501, B*5101, B*5301, CW*0304, CW*0401, and E*0101 is shown

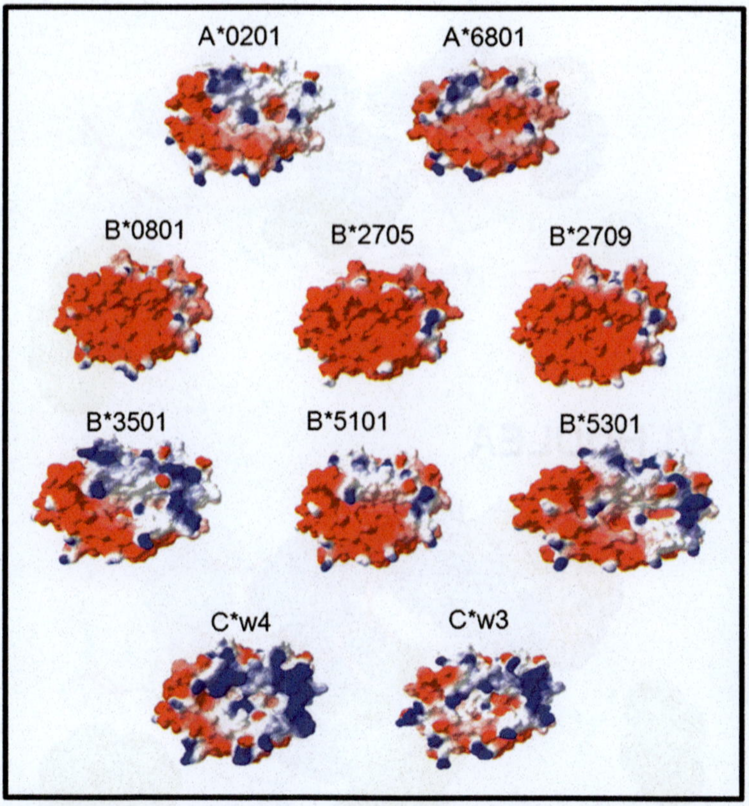

Fig. 15.7 Electrostatic distribution at the peptide-binding grooves of HLA A*0201, A*6801, B*0801, B*2705, B*2709, B*3501, B*5101, B*5301, CW*0304, and CW*0401 is shown. Electropositive distribution is shown in blue, electronegative distribution is shown in red, and neutral distribution is shown in white

15.4 HLA-Peptide Interaction

HLA-peptide interaction is an important step in T-cell-mediated immune response. The binding of two peptides (VLHDDLEA and VLRDDLEA) with HLA*A0201 as shown in Fig. 15.8 is discussed in detail elsewhere using molecular models of HLA-peptide complexes [3]. The consequence of peptide binding caused by single residue mutation as a result of single nucleotide polymorphism (SNP) is demonstrated in this example. Thus, the binding of HLA-peptide binding is combinatorial with reference to HLA alleles and peptide diversity. A comprehensive understanding of HLA-peptide binding is documented using a representative set of HLA-peptide complex structures [4]. The importance of primary and secondary structural anchors at the peptide-binding pockets of HLA molecules is illustrated. The significance of side chain to side chain interaction at the HLA-peptide interface between peptide and HLA is highly intriguing [5]. Databases such as MPID

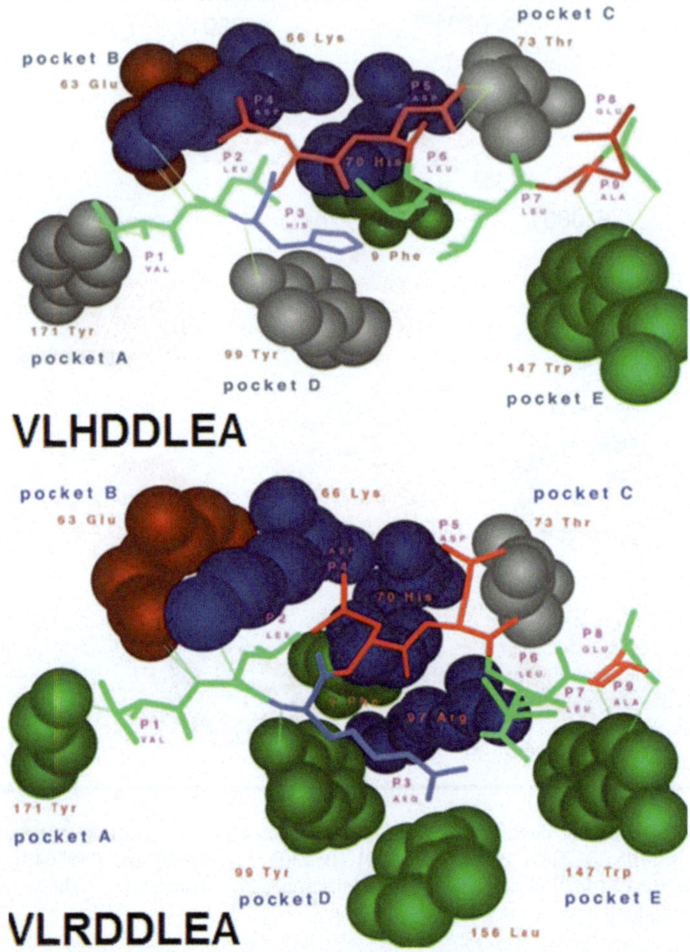

Fig. 15.8 The binding of two peptides having single residue mutation with HLA A*0201 is shown. The mutation is at the third residue where H is replaced by R. The competitive binding IC50 values for first and second peptides are 30 nanomolar and 365 nanomolar, respectively. Thus, the binding of the second peptide with R at the third position is tenfold less strong than the first peptide in binding. This clearly illustrates the consequence of peptide binding caused by single residue mutation and by single nucleotide polymorphism (SNP)

(MHC-Peptide Interaction Database for sequence-structure-function information on peptides binding to MHC molecules) made available for the understanding of HLA-peptide interactions provide ample opportunity to make progress in the field of short peptide vaccine design and development [6]. Structural data on HLA-peptide help to illustrate the lateral view of peptide binding to HLA class I and class II

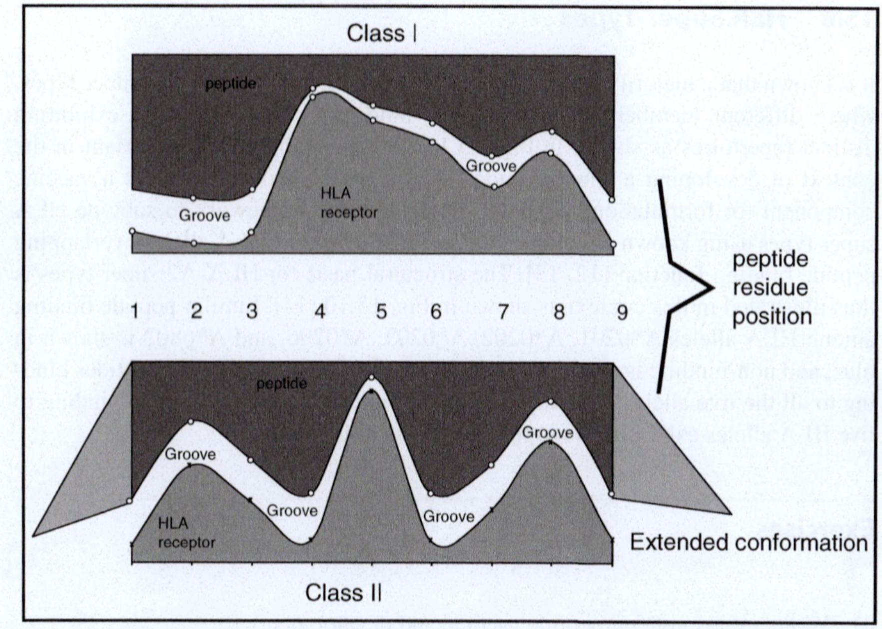

Fig. 15.9 The lateral views of peptide binding to HLA class I and class II molecules are illustrated. This is shown with reference to each peptide residue position. The binding anchor residues are observed at the binding pockets in both class I and class II molecules. It should be noted that class II binding peptides have extended conformation unlike class I binding peptides. It is also interesting to observe that the peptide-binding region in the antigen is nine residues long as shown in this illustration

molecules as shown in Fig. 15.9. The binding anchor residues are observed at the binding pockets in both class I and class II molecules.

15.5 HLA-Peptide Binding Prediction

HLA-peptide binding prediction plays an important role in the design of short peptide antigen *EPITOPES* as vaccine candidates for several infectious and parasitic diseases. The prediction of HLA-peptide binding prediction using structural and modeling principles has been a long-standing interest of our group [7]. We have demonstrated the use of molecular modeling driven vdW-based ranking of modeled HLA-peptide complexes in this context [8]. HLA-peptide prediction using defined structural pockets for both class I [9] and class II [10] HLA alleles with reasonable and reliable efficiency is made available. The subsequent implementation of the model using CGI-driven web server serves the immunology community for short peptide EPITOPE design [11].

15.6 HLA Super-Types

It is known that a majority of HLA alleles are covered within few HLA super-types, where different members of a super-type bind similar peptides, yet exhibiting distinct repertoires as shown in Fig. 15.10. This phenomenon is important in the context of developing a short peptide or a minimal set of peptides as a vaccine component for formulation. We have demonstrated a framework to subtype HLA super-types using known structural information on defined HLA alleles overlapping peptide-binding function [12, 13]. The structural basis for HLA-A2 super-types is thus illustrated in this context as shown in Fig. 15.10 [14]. Similar peptide binding among HLA alleles A*0201, A*0202, A*0203, A*0206, and A*6802 is shown in blue, and non-binding is shown in red. The first block shows a list of peptides binding to all the five alleles demonstrating HLA super-types. Each peptide binding to five HLA alleles exhibiting as super antigens is illustrated.

Exercises

1. Write a short note on peptide vaccine and development.
2. Illustrate T-cell immunity using a neat-labeled diagram.
3. Expand HLA.
4. Expand TCR.
5. What is an allele?
6. State ethnic groups.
7. Illustrate HLA polymorphism.
8. Explain HLA super-type.
9. Illustrate a super antigen.
10. Illustrate the difference in peptide binding with class I and class II HLA molecules.

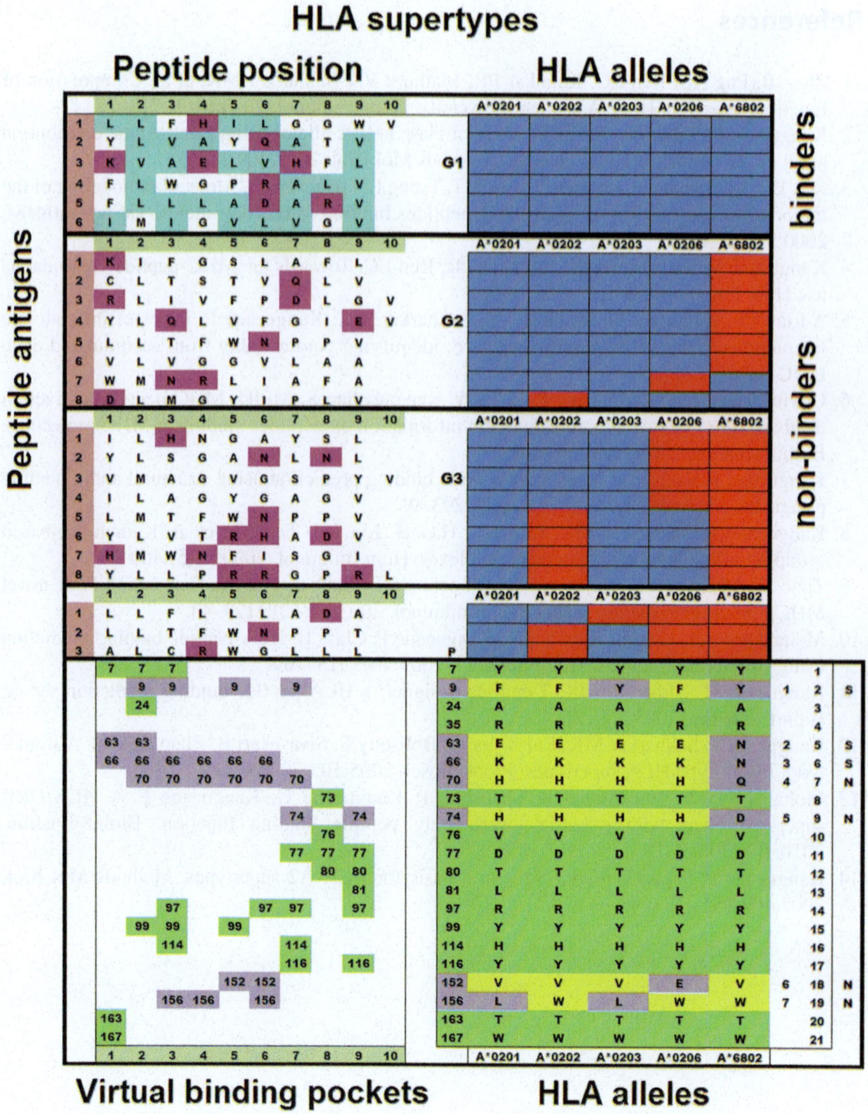

Fig. 15.10 An illustration of HLA A2 super-types is shown. A majority of alleles are covered within few HLA super-types, where different members of a super-type bind similar peptides, yet exhibiting distinct repertoires. Similar peptide binding among HLA alleles A*0201, A*0202, A*0203, A*0206, and A*6802 is shown in blue, and non-binding is shown in red. The first block shows a list of peptides binding to all the five alleles. The second, third, and fourth blocks show peptides binding to more than one alleles. The mutations at the binding pockets of these HLA alleles with reference to peptide-binding residue position are shown in the last block. It should be noted that the mutations are either synonymous or non-synonymous between different alleles in the A2 super-types

References

1. Zhao B, Png AE, Ren EC, Kolatkar PR, Mathura VS, Sakharkar MK, et al. Compression of functional space in HLA-A sequence diversity. Hum Immunol. 2003;64(7):718–28.
2. Kangueane P, Sakharkar MK. Grouping of class I HLA alleles using electrostatic distribution maps of the peptide binding grooves. Methods Mol Biol. 2007;409:175–81.
3. Ren EC, Kangueane P, Kolatkar P, Lin MT, Tseng LH, Hansen JA. Molecular modeling of the minor histocompatibility antigen HA-1 peptides binding to HLA-A alleles. Tissue Antigens. 2000;55(1):24–30.
4. Kangueane P, Sakharkar MK, Kolatkar PR, Ren EC. Towards the MHC-peptide combinatorics. Hum Immunol. 2001;62(5):539–56.
5. Adrian PE, Rajaseger G, Mathura VS, Sakharkar MK, Kangueane P. Types of inter-atomic interactions at the MHC-peptide interface: identifying commonality from accumulated data. BMC Struct Biol. 2002;2:2.
6. Govindarajan KR, Kangueane P, Tan TW, Ranganathan S. MPID: MHC-peptide interaction database for sequence-structure-function information on peptides binding to MHC molecules. Bioinformatics. 2003;19(2):309–10.
7. Kangueane P, Sakharkar MK. HLA-peptide binding prediction using structural and modeling principles. Methods Mol Biol. 2007;409:293–9.
8. Kangueane P, Sakharkar MK, Lim KS, Hao H, Lin K, Chee RE, et al. Knowledge-based grouping of modeled HLA peptide complexes. Hum Immunol. 2000;61(5):460–6.
9. Zhao B, Mathura VS, Rajaseger G, Moochhala S, Sakharkar MK, Kangueane P. A novel MHCp binding prediction model. Hum Immunol. 2003;64(12):1123–43.
10. Mohanapriya A, Lulu S, Kayathri R, Kangueane P, Class II. HLA-peptide binding prediction using structural principles. Hum Immunol. 2009;70(3):159–69.
11. Kangueane P, Sakharkar MK. T-epitope designer: a HLA-peptide binding prediction server. Bioinformation. 2005;1(1):21–4.
12. Kangueane P, Sakharkar MK, Rajaseger G, Bolisetty S, Sivasekari B, Zhao B, et al. A framework to sub-type HLA supertypes. Front Biosci. 2005;10:879–86.
13. Mohanapriya A, Nandagond S, Shapshak P, Kangueane U, Kangueane P. A HLA-DRB supertype chart with potential overlapping peptide binding function. Bioinformation. 2010;4(7):300–9.
14. Kangueane P, Sakharkar MK. Structural basis for HLA-A2 supertypes. Methods Mol Biol. 2007;409:155–62.

Protein-Protein Interfaces and Diseases 16

16.1 Cholera Toxin (CT) Structural Complex

Cholera toxin (CT) is responsible for cholera caused by *Vibrio cholera*. CT is a hetero (different subunit) hexamer (six subunits) consisting of a homo pentamer of five identical CTB subunits and a CTA subunit cleaved into A1 and A2 for functional toxicity as shown in Fig. 16.1. Cholera toxin responsible for cholera is a multimeric protein complex made of six subunits. A stable interface is formed between subunit A and the pentamer of CTB (B_5). Similarly interfaces are formed between each of the five subunits forming the pentamer.

16.2 CTB Interfaces in Cholera Toxin

There are five interfaces in the CTB pentamer complex. Each CTB subunit interacts with two other CTB subunits as shown in Fig. 16.2. Each interface is unique. It is of interest to study each of these two interfaces. The D subunit interacts with the E subunit in the left and the H subunit in the right. The D-E interface is different from D-H interface. Thus, multimeric proteins have different interfaces among themselves between subunits. Analysis of these interfaces will help in the design, development, and production of a stable yet viable subunit-based cholera vaccine.

16.3 Interface Representation in Cholera Toxin

There are three (A/B^5; D-E; D-H) different types of interfaces in a cholera toxin. These interfaces are shown using a linear representation in Fig. 16.3 for a simple understanding. The interface residue positions between CTA and CTB pentamer are shown in Fig. 16.3(a) along with positions corresponding to known mutations among serogroups. This is interesting and will help to understand the functional consequences of subunit-subunit binding by natural mutations in variants. Similarly,

P. Kangueane, C. Nilofer, *Protein-Protein and Domain-Domain Interactions*,
https://doi.org/10.1007/978-981-10-7347-2_16

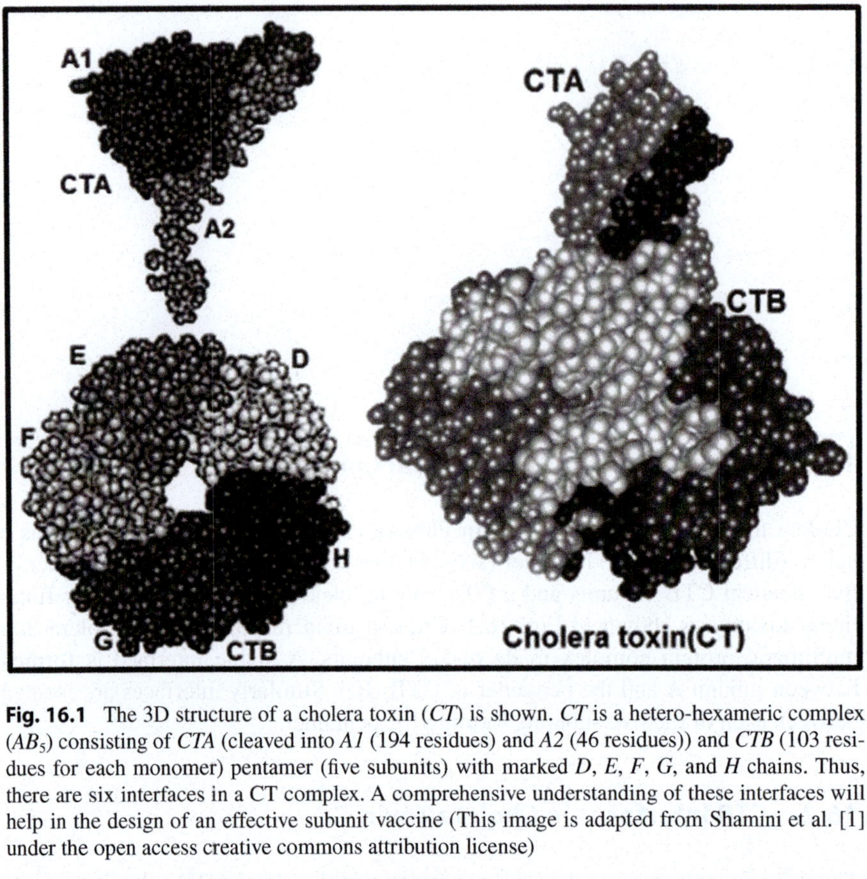

Fig. 16.1 The 3D structure of a cholera toxin (*CT*) is shown. *CT* is a hetero-hexameric complex (AB_5) consisting of *CTA* (cleaved into *A1* (194 residues) and *A2* (46 residues)) and *CTB* (103 residues for each monomer) pentamer (five subunits) with marked *D*, *E*, *F*, *G*, and *H* chains. Thus, there are six interfaces in a CT complex. A comprehensive understanding of these interfaces will help in the design of an effective subunit vaccine (This image is adapted from Shamini et al. [1] under the open access creative commons attribution license)

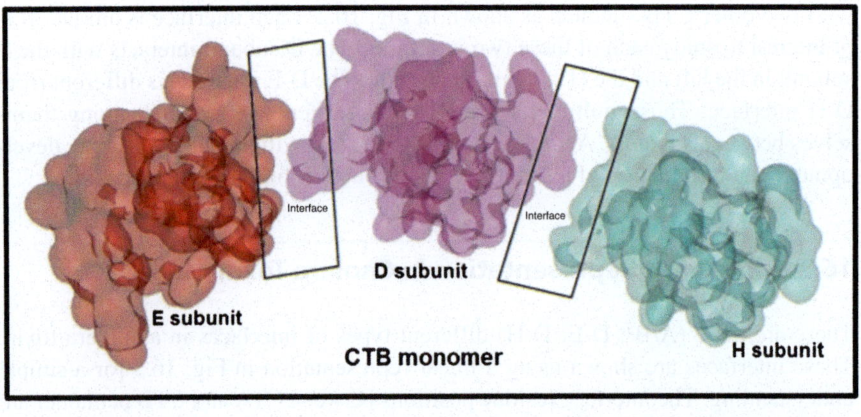

Fig. 16.2 An illustration of a structural model of CTB/CTB interfaces in CTB (B_5) pentamer is shown. Each CTB subunit (D) is juxtaposed by two other CTB units on either side (e.g., E and H). Thus, the D subunit creates two different types of interfaces (D-E and D-H) on either side. This subsequently results in two different interface patterns for subunit D on either side (This image is adapted from Shamini et al. [1] under the open access creative commons attribution license)

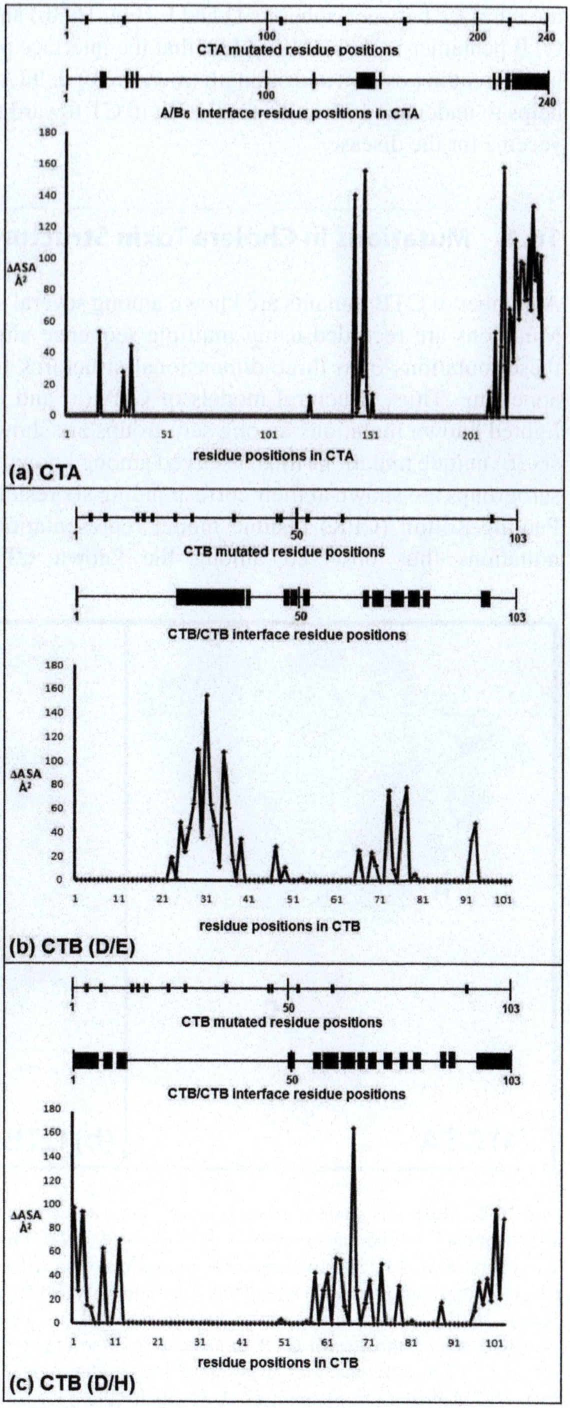

Fig. 16.3 The three different types (A/B_5; D-E; D-H) of interfaces are shown as a measure of delta ASA (change in accessible surface area) against residue positions in a linear plot. Known residue positions of mutations among serogroups are also mapped. A/B_5 interface is given in panel (**a**); D-E interface is given in panel (**b**); and D-H interface is given in panel (**c**) (This image is adapted from Shamini et al. [1] under the open access creative commons attribution license)

the interface between subunits D and E (Fig. 16.3b) and D and H (Fig. 16.3c) of the CTB pentamer is shown. It is clear that the interface pattern between D-E and D-H is different in position and strength (measure of delta ASA). Thus, interface analysis helps to understand complex formation in CT toward the design of a subunit-based vaccine for the disease.

16.4 Mutations in Cholera Toxin Structures

A number of CTB mutants are known among several serogroups of *Vibrio cholera*. Mutations are recorded using multiple sequence alignment (MSA). Mapping of these mutations onto three-dimensional structures for a better understanding is appealing. Thus, structural models of CTA (a) and CTB (b) subunits with highlighted known mutations among serogroups are shown in Fig. 16.4. (a) A total of six (6) unique mutations thus observed among known CTA sequences from several serogroups are shown at their corresponding six residue positions using the Corey-Pauling-Koltun (CPK) residue model representation. (b) Fourteen (14) unique mutations thus observed among the known CTB sequences from several

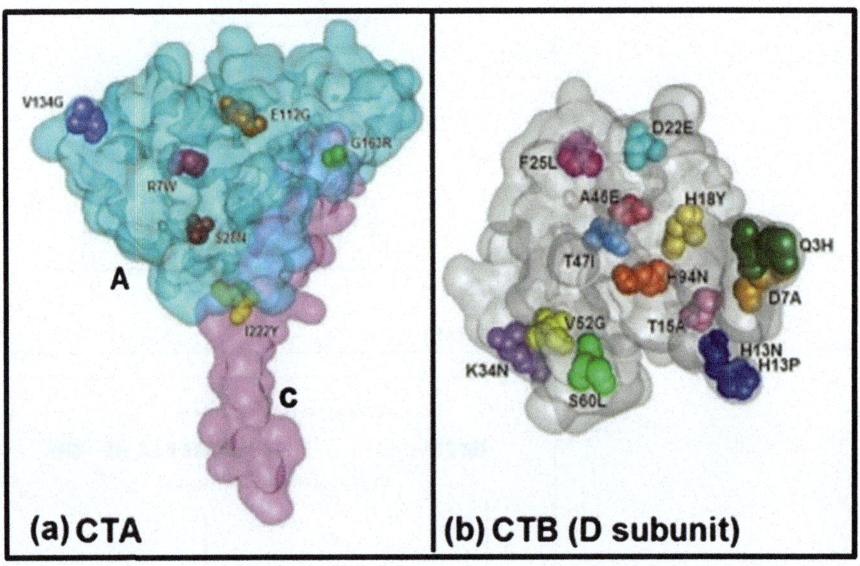

Fig. 16.4 Mutations among serogroups are shown in the structural models of CTA (**a**) and CTB (**b**) subunits. We used the structure with PDB entry (1XTC) for generating this illustration using Discovery studio [3]. Six (6) unique mutations thus observed among known CTA sequences from several serogroups are shown at their corresponding six residue positions using the Corey-Pauling-Koltun (CPK) residue model representation in panel (**a**). Fourteen (14) unique mutations thus observed among the known CTB sequences from several serogroups are shown at their corresponding 13 residue positions using the CPK residue model representation in panel (**b**) (This image is adapted from Shamini et al. [1] under the open access creative commons attribution license)

serogroups are shown at their corresponding 13 residue positions using the CPK residue model representation. This image is adapted from Shamini et al. [1] under the open access creative commons Attribution license.

16.5 Interface Mutations in Cholera Toxin

Mutations are found in cholera toxin among serogroups. Their presence at the subunit-subunit interface is shown in Fig. 16.5. We have marked mutations that are present at the CTA/CTB interface as well as the CTB/CTB interface. This is of importance in the development of a viable subunit vaccine. Thus, interface analysis is crucial in vaccine development.

16.6 HIV-1/GP160 (GP120/GP40) Trimer ENV Spike Complex

The development of an effective HIV-1/AIDS vaccine is an ongoing effort. The consideration of HIV-1/GP160 (cleaved GP120/GP40) trimer ENV spike complex is a biotechnological challenge due to downstream processing issues. The superimposed electron microscopy (EM) structure of HIV-1/GP160 (cleaved GP120/GP40)

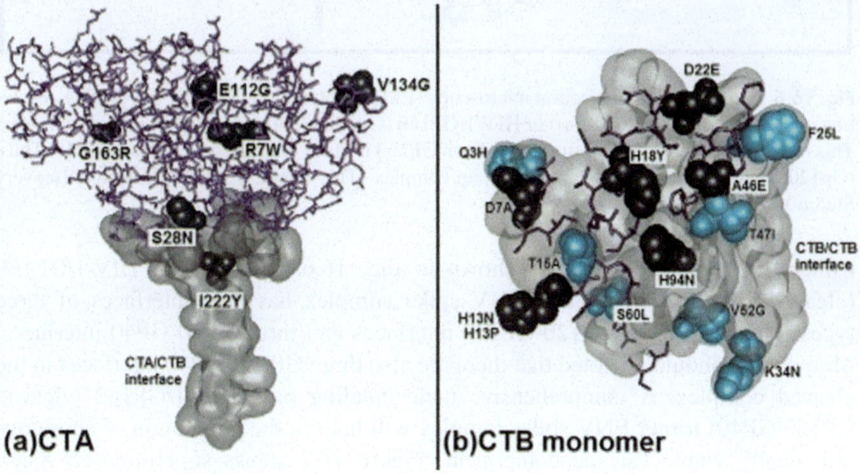

Fig. 16.5 Mutations at the CTB interfaces among serogroups are shown in the structural models of CTA (**a**) and CTB (**b**) subunits with known mutations. We used the structure with PDB entry (1XTC) for generating this visual using the freeware Discovery studio [3]. A total of six unique mutations observed among several serogroups are shown at their corresponding residue positions using the CPK residue model representation. The six mutated positions are present at the solvent-exposed regions of CTA in both monomer and CTA/CTB complex state. A total of 7 out of 14 unique mutations thus observed among the known CTB sequences from several serogroups are shown at their corresponding 7 (3, 15, 25, 34, 47, 52 and 60) out of 14 residue positions using the CPK residue model representation (This image is adapted from Shamini et al. [1] under the open access creative commons attribution license)

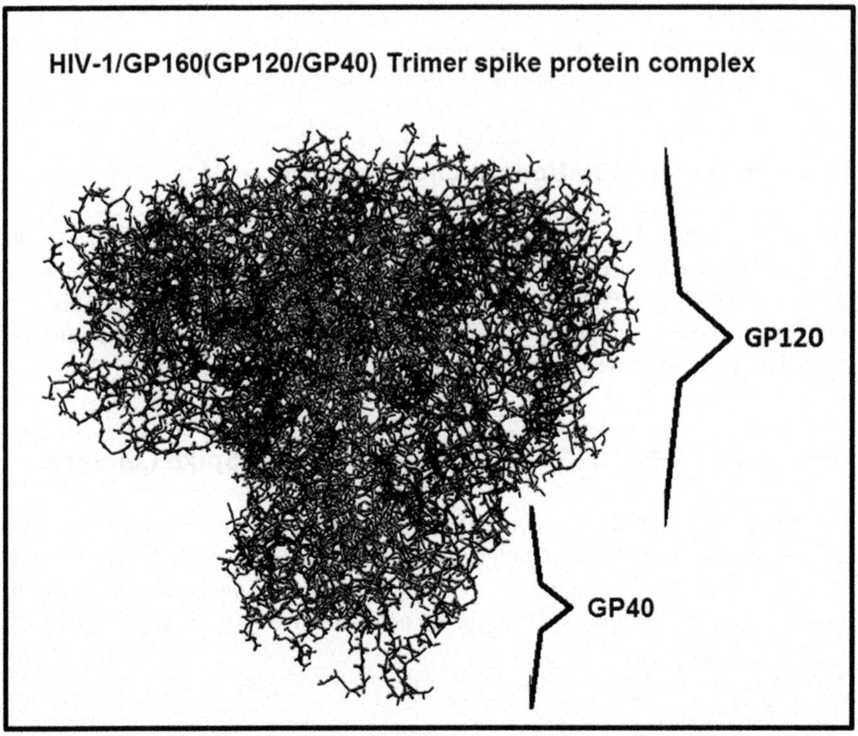

Fig. 16.6 Superposition of electron microscopy (EM) structures (PDB ID: 5FUU (4.19 Å resolution) and 5U1F (6.8 Å resolution)) of HIV-1/GP160 (Gp120/GP40) trimer spike protein complex. This is a trimer of three GP160 structures. Each GP160 is made of cleaved GP120 and GP40. Thus, $(Gp120/GP40)_3$ forms the viral spike protein complex (This image is generated using Discovery Studio Visualizer™ version 16.1 [3])

trimer ENV spike complex is shown in Fig. 16.6. An effective HIV-1/GP160 (cleaved GP120/GP40) trimer ENV spike complex has nine interfaces of three types. There are three GP120-GP120 interfaces and three GP40-GP40 interfaces. Moreover, it should be noted that there are also three GP120-GP40 interfaces in the cleaved complex. A comprehensive understanding of a HIV-1/GP160 (cleaved GP120/GP40) trimer ENV spike complex will help in the production of an immunologically viable vaccine component. Figure 16.7 shows superimposed X-ray structures of GP120 (panel a) and GP40 (panel c). A top view of the electron microscopy structures (EM) of GP120 trimer (Fig. 16.7b) and GP40 trimer (Fig. 16.7d) is also shown. Figure 16.8 shows residue position specific ASA distribution profiles for (a) gp120 and (b) gp41, (c) and (d) respective means, and (e) and (f) respective standard deviations. These observations provide valuable insight in the context of understanding the structural conformation of HIV-1/GP160 (cleaved GP120/GP40) trimer ENV spike complex.

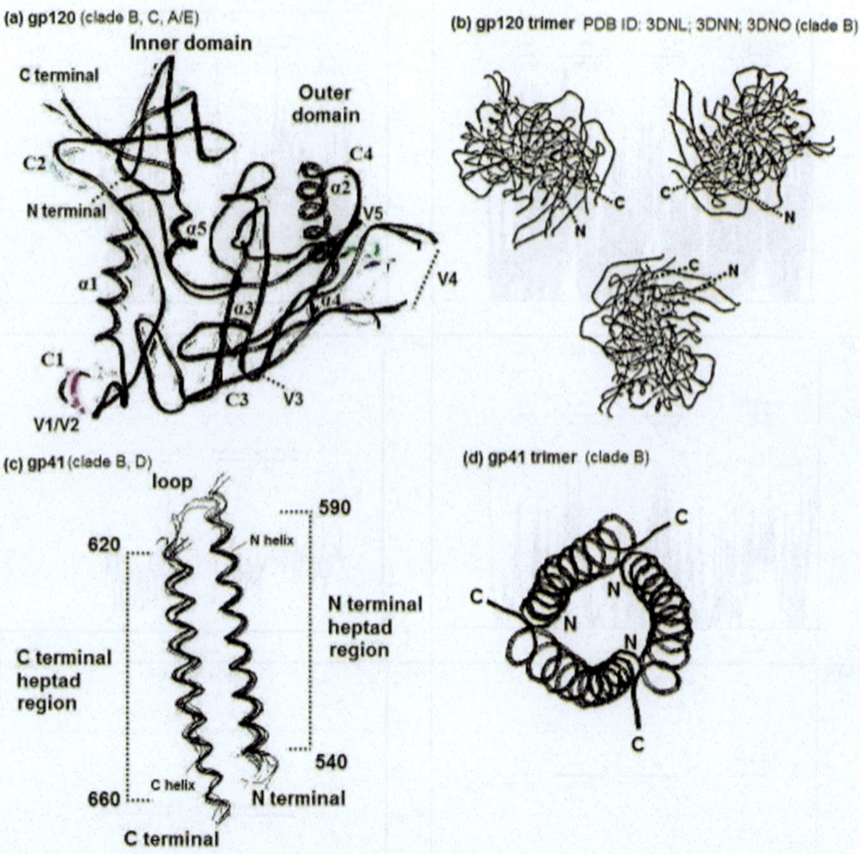

Fig. 16.7 Structural superimposition of (**a**) gp120, (**b**) gp120 trimer, (**c**) gp41, and (**d**) gp41 trimer is shown using structures obtained from PDB. The structures were superimposed using the software SPDBV (Swiss PDB Viewer version 3.7). The variable loops (V1-V5) and the constant regions (C1-C4) of the GP120 structure, with inner and outer domains, are shown (**a**). The superimposition of the gp120 trimer (PDB ID: 3DNL), solved through electron microscopy at 20.0 Å, is shown in (**b**). The N heptad (a repeating seven-residue pattern) and the C heptad (a repeating seven-residue pattern) regions of the superimposed gp41 structure, solved by X-ray crystallography at a resolution of 2.10 Å, are shown in (**c**). Only the N heptad region of gp41 is available in trimeric form, as illustrated in (**d**) (This image is adapted from Sowmya et al. [2] under the open access creative commons attribution license)

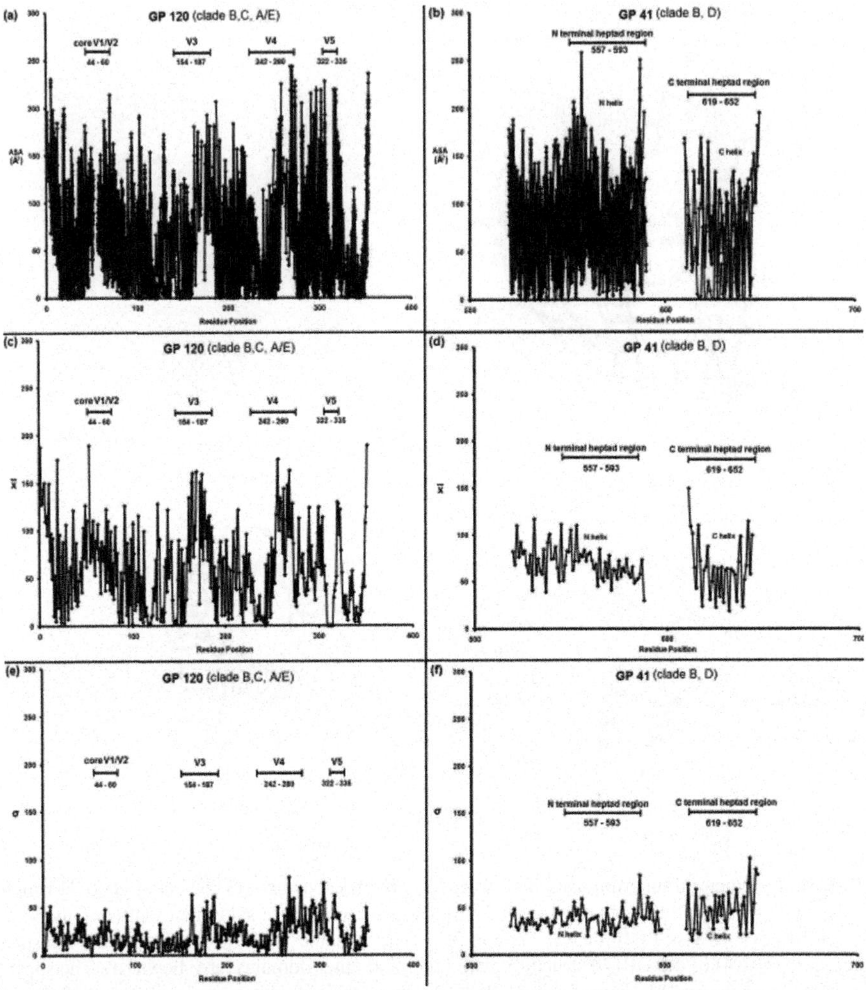

Fig. 16.8 Residue position-specific ASA distribution profiles for (**a**) gp120 and (**b**) gp41, (**c**) and (**d**) respective means, and (**e**) and (**f**) respective standard deviations are shown using structures obtained from PDB. The ASA distribution was calculated using surface racer [4] (This image is adapted from Sowmya et al. [2] under the open access creative commons attribution license)

Exercises

1. Illustrate the structure of a functional cholera toxin.
2. How many subunits constitute a cholera toxin?
3. Illustrate the HIV-1 spike protein using a neat diagram.
4. How many subunits constitute HIV-1/ENV GP160 trimer?
5. Discuss the issues in the development of an effective cholera vaccine.
6. Discuss the issues in the development of an effective HIV-1/AIDS vaccine.
7. Illustrate the CTB and CTA interface in a cholera toxin.
8. Illustrate the CTB interfaces in a cholera toxin.
9. Illustrate the GP120/GP40 interface in the HIV-1/ENV protein.
10. Discuss HIV-1/ENV GP160 trimer spike structure.

References

1. Shamini G, Ravichandran M, Sinnott JT, Somboonwit C, Sidhu HS, Shapshak P, Kangueane P. Structural inferences for cholera toxin mutations in vibrio cholerae. Bioinformation. 2011;6(1):1–9.
2. Sowmya G, Shamini G, Anita S, Sakharkar M, Mathura V, Rodriguez H, Levine AJ, Singer E, Commins D, Somboonwit C, Sinnott JT, Sidhu HS, Rajaseger G, Pushparaj PN, Kangueane P, Shapshak P. HIV-1 envelope accessible surface and polarity: clade, blood, and brain. Bioinformation. 2011;6(2):48–56.
3. Dassault Systemes BIOVIA, BIOVIA Discovery Studio Visualizer, v16.1.0.15350, San Diego: Dassault Systemes, 2015 [cited: 2017 Mar 20]. Available from: http://accelrys.com/products/collaborative-science/biovia-discovery-studio/visualization-download.php
4. Tsodikov OV, Record MT Jr, Sergeev YV. Novel computer program for fast exact calculation of accessible and molecular surface areas and average surface curvature. J Comp Chem. 2002;23:600–9.

Patented Protein Structural Complexes in Discovery Platform

17

17.1 Patented Protein Complexes

Protein-protein complexes associated with several diseases have been considered as drug targets in recent years. A number of such complexes have been patented by many organizations as shown in Table 17.1. The procedure for the selection of this list of complexes is shown in Fig. 17.1. A number of such complexes have known 3D structures determined by X-ray crystallography as shown in Figs. 17.2, 17.3, 17.4, 17.5, 17.6, 17.7, 17.8, 17.9, 17.10, and 17.11. Table 17.1 gives a list of gene products like BCAT, KCAT, and others associated with specific diseases. Information related to patent ID, GenBank ID, date of patent, disease association, and target and known structures are also reported. Thus, it is clear that there is interest in considering protein-protein complexes as drug targets in addition to classical approach of selecting monomer proteins as drug targets.

17.2 Selection of Patented Protein Complex Structures

We used GenPept at the National Center for Biotechnology Information (NCBI) to search for records with disease-associated known protein complex structures having patent rights as shown in Fig. 17.1. This information is summarized in Table 17.1.

17.3 Breast Cancer-Associated BCAT Complexes

A number of protein complexes homologous to the human breast cancer-associated BCAT are shown in Figs. 17.2, 17.3, 17.4, 17.5, and 17.6. The structural representation of these structures provides valuable insights in this context.

P. Kangueane, C. Nilofer, *Protein-Protein and Domain-Domain Interactions*,
https://doi.org/10.1007/978-981-10-7347-2_17

Table 17.1 List of patented protein complexes with known structures (PDB entry) that are associated with diseases

Genbank ID	Patent ID	Date	Target	Disease	Country	PDB ID
ADT50586	7842467-B1 2603	13/Dec/2010	BCAT	Breast cancer	Celera Corporation; Alameda, CA; USA	2EK1
ADT50585	7842467-B1 2602	13/Dec/2010	BCAT	Breast cancer	Celera Corporation; Alameda, CA; USA	2EK1
ADT50584	7842467-B1 2601	13/Dec/2010	BCAT	Breast cancer	Celera Corporation; Alameda, CA; USA	2EK1
ADT50583	7842467-B1 2600	13/Dec/2010	BCAT	Breast cancer	Celera Corporation; Alameda, CA; USA	2EK1
ADT51134	7842467-B1 3151	13/Dec/2010	BCAT	Breast cancer	Celera Corporation; Alameda, CA; USA	5UO7
ADT51133	7842467-B1 3150	13/Dec/2010	BCAT	Breast cancer	Celera Corporation; Alameda, CA; USA	5UO7
ADT50452	7842467-B1 2469	13/Dec/2010	BCAT	Breast cancer	Celera Corporation; Alameda, CA; USA	1ZBQ
ADT50451	7842467-B1 2468	13/Dec/2010	BCAT	Breast cancer	Celera Corporation; Alameda, CA; USA	1ZBQ
ADT50450	7842467-B1 2467	13/Dec/2010	BCAT	Breast cancer	Celera Corporation; Alameda, CA; USA	1ZBQ
ADT50045	7842467-B1 2062	13/Dec/2010	BCAT	Breast cancer	Celera Corporation; Alameda, CA; USA	4AH6
ADT50044	7842467-B1 2061	13/Dec/2010	BCAT	Breast cancer	Celera Corporation; Alameda, CA; USA	4AH6
ADT49526	7842467-B1 1543	13/Dec/2010	BCAT	Breast cancer	Celera Corporation; Alameda, CA; USA	2QUH
ADT46714	7842466-B1 375	13/Dec/2010	CCAT	Colon	Celera Corporation; Alameda, CA; USA	2QUH
ADT48683	7842467-B1 700	13/Dec/2010	BCAT	Breast cancer	Celera Corporation; Alameda, CA; USA	4J15
ADT48682	7842467-B1 699	13/Dec/2010	BCAT	Breast cancer	Celera Corporation; Alameda, CA; USA	4J15
ADT48681	7842467-B1 698	13/Dec/2010	BCAT	Breast cancer	Celera Corporation; Alameda, CA; USA	4J15
ADT48680	7842467-B1 697	13/Dec/2010	BCAT	Breast cancer	Celera Corporation; Alameda, CA; USA	4J15
ADT48679	7842467-B1 696	13/Dec/2010	BCAT	Breast cancer	Celera Corporation; Alameda, CA; USA	4J15
ADA20057	7608413-A 1881	14/Dec/2009	KCAT	Kidney	Celera Corporation; Alameda, CA; USA	4ZYA
ADA20056	7608413-A 1880	14/Dec/2009	KCAT	Kidney	Celera Corporation; Alameda, CA; USA	4ZYA
ADA20055	7608413-A 1879	14/Dec/2009	KCAT	Kidney	Celera Corporation; Alameda, CA; USA	4ZYA
ACM81016	6812339-A 6514	10/Feb/2009	SNPs	Human	Applera Corporation; Norwalk, CT, USA	4ZYA
ADA20228	7608413-A 2052	14/Dec/2009	KCAT	Kidney	Celera Corporation; Alameda, CA; USA	2QUH
ADA19982	7608413-A 1806	14/Dec/2009	KCAT	Kidney	Celera Corporation; Alameda, CA; USA	4AH6
ADA19981	7608413-A 1805	14/Dec/2009	KCAT	Kidney	Celera Corporation; Alameda, CA; USA	4AH6
AKX87644	8999380-B2 995	13/Aug/2015	transferase	Transferase related	Moderna Therapeutics, Inc.; Cambridge, MA, USA	5IN3
ACM80472	6812339-A 5970	10/Feb/2009	synthase	Neo-plastic	Applera Corporation; Norwalk, CT	2LL6
AAB78616	5658565-A 2	8/Oct/1997	synthase	Neo-plastic		2LL6
AGV80639	8524218-B2 66	17/Sep/2013	synthase	Neo-plastic	Pebble Beach CA; USA	4UX6
ADS49384	7803637-B2 66	13/Dec/2010	synthase	Neo-plastic	Jenomic; Carmel, CA; USA	4UX6
ADS26861	7777097-B2 730	12/Dec/2010	-	Plant disease resistance genes	Syngenta Participations AG; Basel; CH, SWIZERLAND	Unknown
ADT49035	7842467-B1 1052	13/Dec/2010	BCAT	Breast cancer	Celera Corporation; Alameda, CA; USA	Unknown
ADT49034	7842467-B1 1051	13/Dec/2010	BCAT	Breast cancer	Celera Corporation; Alameda, CA; USA	Unknown
ADT49033	7842467-B1 1050	13/Dec/2010	BCAT	Breast cancer	Celera Corporation; Alameda, CA; USA	Unknown

PDB = protein databank

PDB **protein databank**

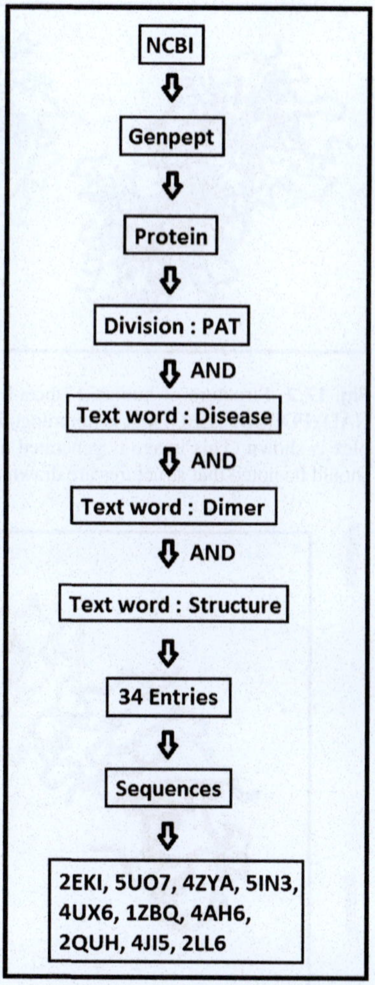

Fig. 17.1 A flowchart for the identification of patented protein complexes associated with disease

17.4 Kidney Disease-Associated KCAT Complexes

A number of protein complexes homologous to the human kidney disease-associated KCAT are shown in Figs. 17.7 and 17.8. The structural representation of these structures provides valuable insights in this context.

17.5 Inducible Nitric Oxide Synthase (iNOS)

A number of protein complexes homologous to the human neoplastic gene-related disease are shown in Figs. 17.9 and 17.10. The structural representation of these structures provides valuable insights in this context.

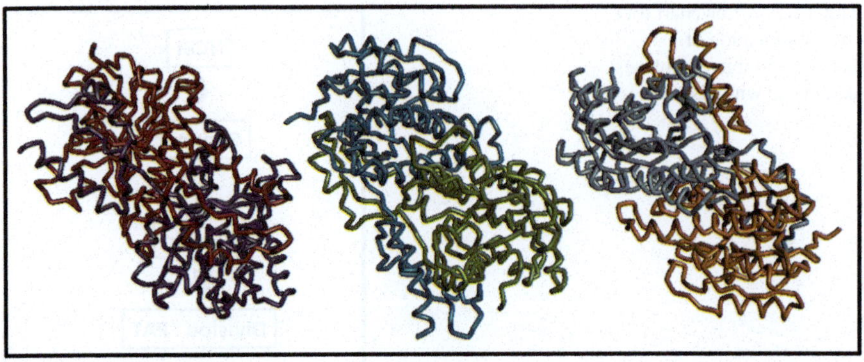

Fig. 17.2 Structure of human 17-beta-hydroxysteroid de-hydrogenase type 4 in complex with NAD (PDB ID: 1ZBQ) that is homologous to the human breast cancer BCAT protein homo complex is shown (This image is generated using Discovery Studio Visualizer™ version 16.1 [1]. It should be noted that structures are drawn as data is available in the PDB entry)

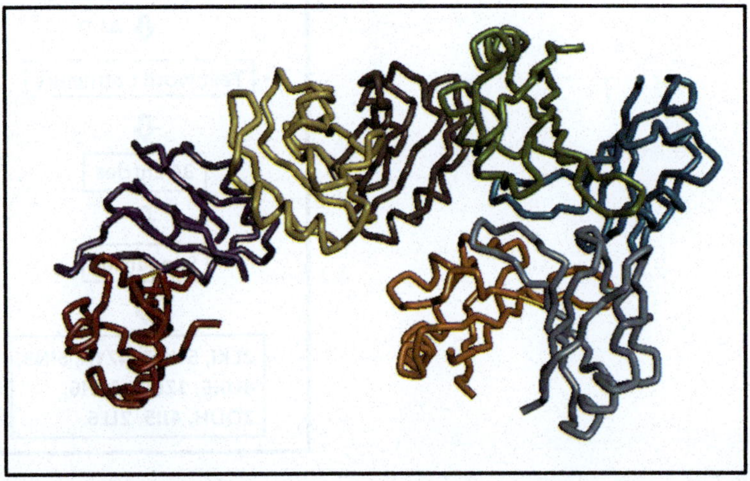

Fig. 17.3 Structure of RNA-binding motif of human RNA-binding protein 12 (PDB ID: 2EK1) that is homologous to the human breast cancer BCAT protein homo complex is shown (This image is generated using Discovery Studio Visualizer™ version 16.1 [1])

17.6 Human Galactose-1-Phosphate Uridylyl-Transferase

The protein complex homologous to the human disease-related target is shown in Fig. 17.11. The structural representation of the structure provides valuable insights in this context.

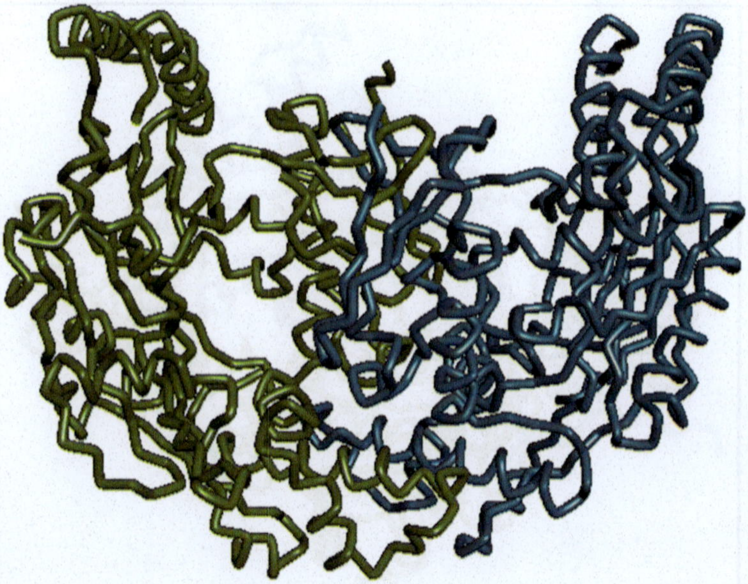

Fig. 17.4 Structure of human neuronal nitric oxide synthase heme domain in complex with (S)-3-[(2-amino-4-methylquinolin-7-yl) methoxy]-5-(2-(methyl amino) propyl) benzonitrile (PDB ID: 5UO7) that is homologous to the human breast cancer BCAT homodimer protein complex is shown (This image is generated using Discovery Studio Visualizer™ version 16.1 [1])

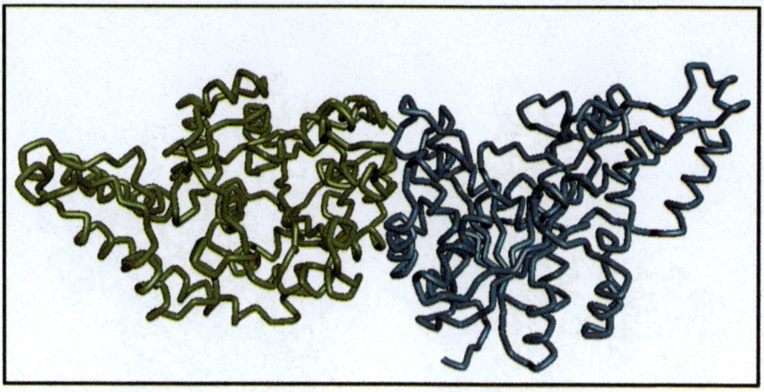

Fig. 17.5 Structure of human tryptophanyl-tRNA synthetase in complex with Trp (PDB ID: 2QUH) that is homologous to the human breast or colon cancer BCAT protein homo complex is shown (This image is generated using Discovery Studio Visualizer™ version 16.1 [1])

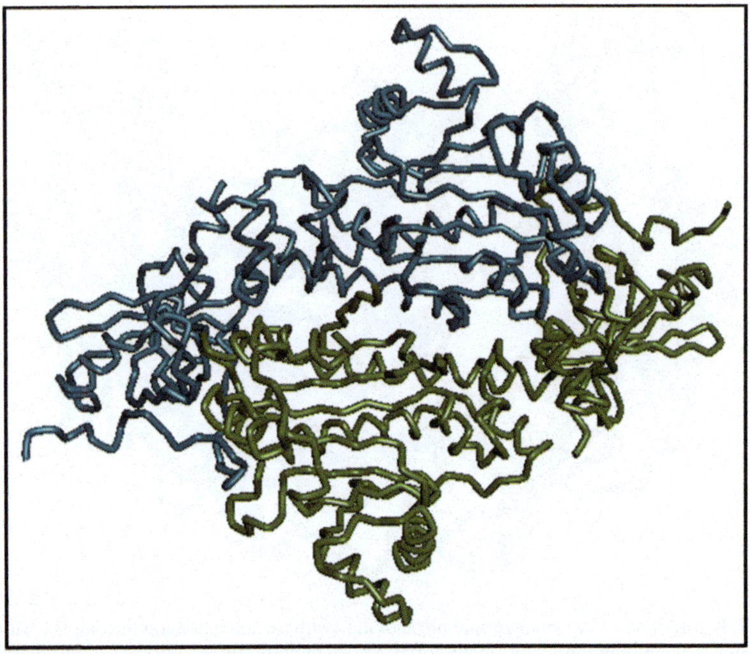

Fig. 17.6 Structure of human cytosolic aspartyl-tRNA synthetase, a component of multi-tRNA synthetase complex (PDB ID: 4J15) that is homologous to the human breast cancer-related BCAT homodimer protein complex, is shown (This image is generated using Discovery Studio Visualizer™ version 16.1 [1])

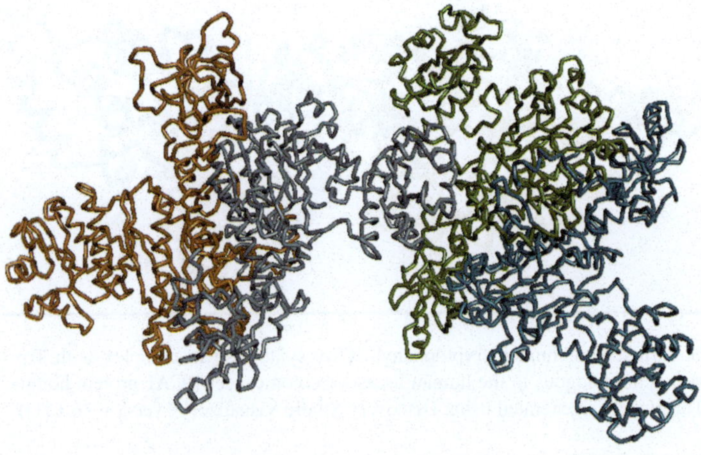

Fig. 17.7 Human mitochondrial aspartyl-tRNA synthetase (PDB ID: 4AH6) that is homologous to the human KCAT homo complex associated with kidney disease is shown (This image is generated using Discovery Studio Visualizer™ version 16.1 [1])

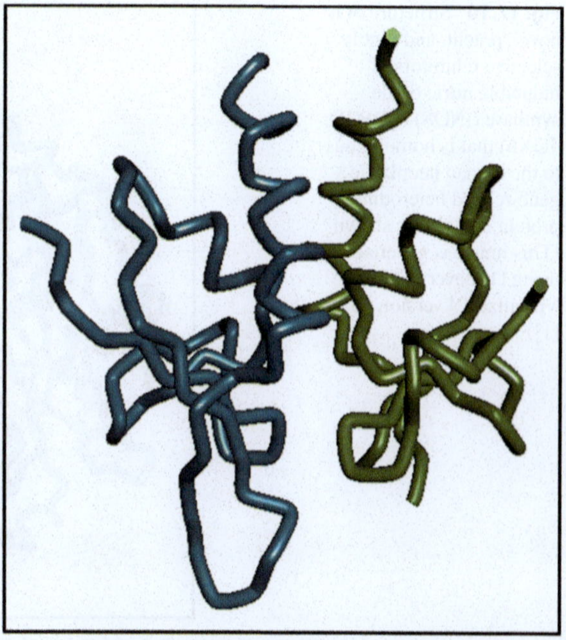

Fig. 17.8 The N-terminal extension domain of human asparaginyl-tRNA synthetase (PDB ID: 4ZYA) that is homologous to human kidney-related KCAT homodimer protein complex is shown (This image is generated using Discovery Studio Visualizer™ version 16.1 [1])

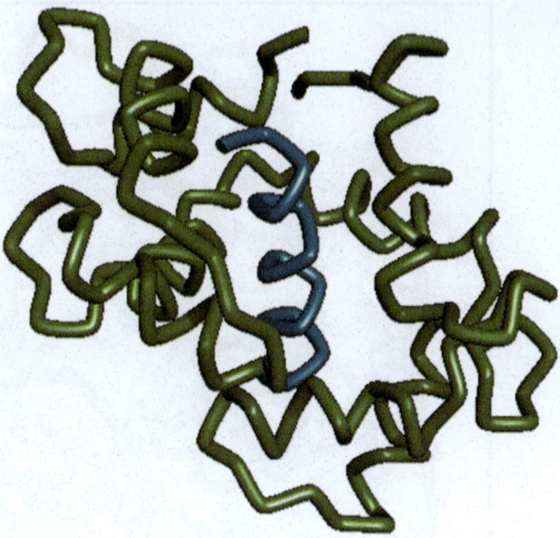

Fig. 17.9 Structure of CaM bound to human inducible nitric oxide synthase (iNOS) CaM-binding domain peptide (PDB ID: 2LL6) is shown (This image is generated using Discovery Studio Visualizer™ version 16.1 [1])

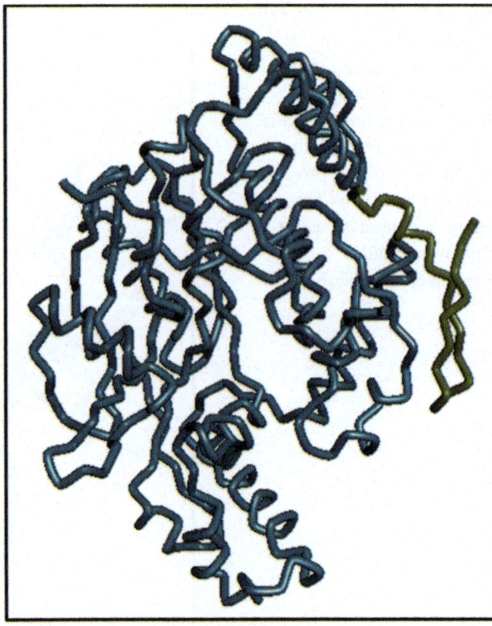

Fig. 17.10 Structure of a novel, potent, and highly selective inhibitors of inducible nitric oxide synthase (iNOS) (PDB ID: 4UX6) that is homologous to the human neoplastic gene-related heterodimer protein complex is shown (This image is generated using Discovery Studio Visualizer™ version 16.1 [1])

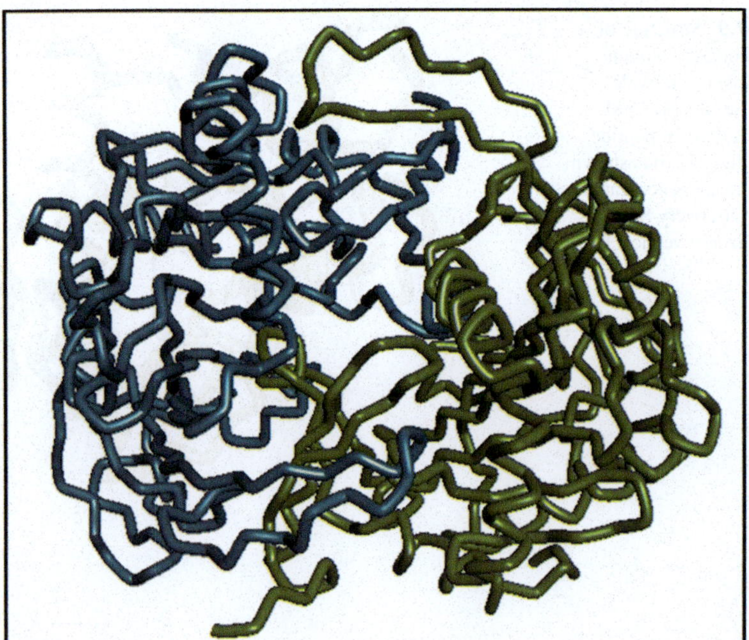

Fig. 17.11 Structure of glucose-1-phosphate-bound nucleotidylated human galactose-1-phosphate uridylyl-transferase (PDB ID: 5IN3) that is homologous to the human homodimer protein complex is shown (This image is generated using Discovery Studio Visualizer™ version 16.1 [1])

Exercises

1. What is a patent?
2. Expand IPR.
3. Illustrate a homodimer.
4. Illustrate a heterodimer.
5. Give some examples of protein complexes as cancer targets.
6. State organizations that have patented protein complexes as drug targets.
7. Expand PAT in GenBank divisions.
8. Give a flowchart to identify known dimer structures that are patented.
9. Describe structures of complexes associated with diseases.
10. Discuss BCAT and KCAT.

Reference

1. Dassault Systemes BIOVIA, BIOVIA Discovery Studio Visualizer, v16.1.0.15350, San Diego: Dassault Systemes, 2015 [cited: 2017 Mar 20]. Available from: http://accelrys.com/products/collaborative-science/biovia-discovery-studio/visualization-download.php

Proteome-Scale Analysis of Protein Complexes

18

18.1 Protein Complexes in Proteomes

A complete set of proteins and their complexes if any in a species organism coded by its genome constitute a proteome. The number of protein complexes in each species genome is alarming in number. Therefore, it is of interest to understand protein-protein interactions in a proteome. A PubMed search at the National Centre for Biotechnology Information (NCBI) for terms "protein-protein" and "proteome" at the title record with some manual curation listed 16 records as on July 22, 2017 [1–16]. Thus, a chronological understanding of this phenomenon is compelling. Proteome-level understating of protein-protein interaction is nontrivial and complex to realize. This requires robust and efficient techniques for such massive analysis and interpretation towards a reasonable approximation. Moreover, the need to understand such phenomenon at a proteome level is systemic in nature. This will lead to the description of their molecular mechanism in a complex cellular environment in future investigations. These attempts have been reviewed in detail as described elsewhere [10, 14]. Molecular mechanics based geometric optimization aided by chemical characterization is the most common method in proteome-scale description of protein-protein interactions [2, 7–9, 11, 12, 16]. Yeast two-hybrid systems [3–5] and peptide arrays [1, 6] were also traditionally used in establishing protein-protein networks. These methods were applied in model systems such as *E. coli* [6], human [4, 5], yeast [2, 3], *Trypanosoma* [8], cyanobacterium *Spirulina platensis* [13], and cancer proteome [15] for proteome-scale understanding of protein networks. We expect more clarity in the future with the availability of additional data in the proteome space.

P. Kangueane, C. Nilofer, *Protein-Protein and Domain-Domain Interactions*,
https://doi.org/10.1007/978-981-10-7347-2_18

Exercises

1. State the number and names of completely sequenced genomes.
2. State the number of genes in each completely sequenced genomes.
3. Enumerate the proteome size in completely sequenced genomes.
4. Illustrate the relationship between proteome size and protein networks.
5. What are the techniques used to study protein-protein interactions at a proteome scale?

References

1. Bialek K, Swistowski A, Frank R. Epitope-targeted proteome analysis: towards a large-scale automated protein-protein-interaction mapping utilizing synthetic peptide arrays. Anal Bioanal Chem. 2003;376(7):1006–13.
2. Lu L, Arakaki AK, Lu H, Skolnick J. Multimeric threading-based prediction of protein-protein interactions on a genomic scale: application to the Saccharomyces cerevisiae proteome. Genome Res. 2003;13(6A):1146–54.
3. Grigoriev A. On the number of protein-protein interactions in the yeast proteome. Nucleic Acids Res. 2003;31(14):4157–61.
4. Stelzl U, Worm U, Lalowski M, Haenig C, Brembeck FH, Goehler H, Stroedicke M, et al. A human protein-protein interaction network: a resource for annotating the proteome. Cell. 2005;122(6):957–68.
5. Rual JF, Venkatesan K, Hao T, Hirozane-Kishikawa T, Dricot A, Li N, et al. Towards a proteome-scale map of the human protein-protein interaction network. Nature. 2005;437(7062):1173–8.
6. Oishi Y, Yunomura S, Kawahashi Y, Doi N, Takashima H, Baba T, et al. Escherichia coli proteome chips for detecting protein-protein interactions. Proteomics. 2006;6(24):6433–6.
7. Wang H, Segal E, Ben-Hur A, Li QR, Vidal M, Koller D. InSite: a computational method for identifying protein-protein interaction binding sites on a proteome-wide scale. Genome Biol. 2007;8(9):R192.
8. Rodriguez-Soca Y, Munteanu CR, Dorado J, Pazos A, Prado-Prado FJ, González-Díaz H. Trypano-PPI: a web server for prediction of unique targets in trypanosome proteome by using electrostatic parameters of protein-protein interactions. J Proteome Res. 2010;9(2):1182–90.
9. Tuncbag N, Gursoy A, Nussinov R, Keskin O. Predicting protein-protein interactions on a proteome scale by matching evolutionary and structural similarities at interfaces using PRISM. Nat Protoc. 2011;6(9):1341–54.
10. Liu ZP, Chen L. Proteome-wide prediction of protein-protein interactions from high-throughput data. Protein Cell. 2012;3(7):508–20. https://doi.org/10.1007/s13238-012-2945-1.
11. Xu Y, Wang H, Nussinov R, Ma B. Protein charge and mass contribute to the spatio-temporal dynamics of protein-protein interactions in a minimal proteome. Proteomics. 2013;13(8):1339–51.
12. Kuzu G, Gursoy A, Nussinov R, Keskin O. Exploiting conformational ensembles in modeling protein-protein interactions on the proteome scale. J Proteome Res. 2013;12(6):2641–53.
13. Senachak J, Cheevadhanarak S, Hongsthong A. SpirPro: a Spirulina proteome database and web-based tools for the analysis of protein-protein interactions at the metabolic level in Spirulina (Arthrospira) platensis C1. BMC Bioinform. 2015;16:233.

14. Keskin O, Tuncbag N, Gursoy A. Predicting protein-protein interactions from the molecular to the proteome level. Chem Rev. 2016;116(8):4884–909.
15. Xu D, Jalal SI, Sledge GW, Meroueh SO. Small-molecule binding sites to explore protein-protein interactions in the cancer proteome. Mol BioSyst. 2016;2(10):3067–87.
16. Maheshwari S, Brylinski M. Across-proteome modeling of dimer structures for the bottom-up assembly of protein-protein interaction networks. BMC Bioinform. 2017;18(1):257.

19 Conclusions and Challenges

This book introduces concepts in protein-protein binding to undergraduate, graduate, and research students in a step-by-step manner using examples, illustrations, and exercises for an easy understanding of this phenomenon. It illustrates the importance and significance of the molecular (physical and chemical) and evolutionary (gene fusion) principles of protein-protein and domain-domain interactions toward the understanding of cell division, disease mechanism, and target definition in drug discovery. The chapters help to glean information from data pertaining to the types of protein-protein complexes (homodimer, heterodimer and multimer protein complexes) in the context of various specific biological functions such as catalysis, regulation, biological assembly, immune response, and inhibition. The significance of such complex formation in biology in the light of molecular evolution is also highlighted using relevant examples.

Protein-protein interaction is an important phenomenon in molecular and cellular biology. A number of molecular events such as catalysis, regulation, transport, signaling, immunity, and inhibition occur through protein-protein interaction in several species organism. It is also found that molecular events that are assisted through protein-protein interactions in some species organisms are guided through domain-domain interactions, in some other species organisms using evolved protein architectures through described gene fusions resulting in naturally occurring chimeric proteins. The mechanisms of such events are unclear in molecular evolution. However, the hypothesis for describing such observation guided by biochemical molecular pressure is gaining clarity with increased data.

The binding of proteins occurs in different forms. Two identical proteins bind with one another forming homodimers. Similarly, two non-homologous proteins bind with one another forming heterodimers. These proteins do bind in several combinations forming multimeric protein complexes consisting of homologous and non-homologous proteins. Thousands of such interactions occur in nature. A number of complexes are also associated with several known diseases and are considered as drug and vaccine candidates. Therefore, it is important to understand the molecular principles of protein-protein binding.

P. Kangueane, C. Nilofer, *Protein-Protein and Domain-Domain Interactions*,
https://doi.org/10.1007/978-981-10-7347-2_19

The molecular principles of protein-protein binding are often deciphered using known three-dimensional (3D) structures of protein complexes solved by X-ray, nuclear magnetic resonance (NMR), and electron microscopy. 3D coordinate data for thousands of such complexes are now available at the protein databank (PDB). This provides an opportunity to describe protein-protein interfaces in a variety of complexes to glean common features. The binding of two proteins happens by their interactions forming stable interfaces. Large interfaces (interface size) have large interacting interface area and hence high binding energy.

The total binding energy between proteins is often described as a sum of van der Waals (vdW), hydrogen bonds (H bonds), and electrostatics at an atomic level. vdW is known to be dominant at the interfaces, and it is also proportionate to interface size irrespective of molecular function. It is also found that H bonds are proportionate to interface size in obligatory complexes such as enzymes, regulators, and biological assembly. However, electrostatics is proportionate in non-obligatory regulator inhibitor complexes. Thus, protein-protein interfaces are vdW dominant with selective hydrogen bonds and or electrostatics toward broad molecular function.

Residue-level analysis of the interfaces shows that interfaces are rich in hydrophobic residues in majority of complexes. However, there are a considerable number of complexes that are not rich in hydrophobic residues. Nonetheless, the core regions are always rich in hydrophobic residues unlike surface regions with rich polar residues. It should be noted that the interface is that part of the surface in a monomer state of the proteins that is involved in protein-protein binding. Thus, characteristics of the interface are distinctly different from the rest of the surface. Therefore, it is important to describe the driving force for protein-protein interaction (PPI). A combined description of interfaces using residue (hydrophobic and hydrophilic) features with atomic characteristics (vdW, H bonds and electrostatics) will provide more clarity on the driving force for protein-protein binding. We foresee more insight in future with increased data.

Protein-protein binding is interesting. However, homodimer binding is fascinating. Furthermore, homodimer folding is intriguing in nature. The description of homodimer interfaces based on folding mechanism is captivating. The presence of 2-state (2S) homodimers without folding intermediates with high interface area is startling. Similarly, the knowledge on the structural interfaces of 3-state intermediates with monomer intermediate (3SMI) and dimer intermediate (3SDI) is adding value to the understanding of protein-protein interactions.

A number of protein-protein complexes are associated with disease progression and are considered as targets for drug and vaccine development. Therefore, it is important to understand protein-protein interaction in these disease-linked complexes. We illustrated and discussed interfaces in a hexameric (six subunits) cholera toxin (CT) and a HIV-1/ENV GP160 (GP120/GP40) trimer (three subunits) spike protein complex. Thus, interface analysis helps in the development of improved vaccine candidates in the prevention of diseases.

Molecular functions are conserved in evolution, and it is deterministic of structurally viable interfaces in several cases. There is a need to quantify functional consequences caused by genetic variation, mutation, and evolution. A precise understanding of protein-protein interaction will help in mimicking and design of stable interfaces taking into consideration mutations across known serogroups. Moreover, the mechanism and hypothesis to describe gene fusion for conserved functions using evolved structural interfaces are noteworthy. The principles of protein-protein interactions in the context of gene fusion leading to domain-domain interfaces are persuasive in nature.

Millions of new sequences are archived in databases every year through genomics, functional genomics, proteomics, and clinical science projects. This will become a routine, and its data size will be alarming in the coming years. Functional annotations of these sequences are always tedious in nature. This is often difficult, less converging, frequently divergent, less comprehensive, and usually incomplete. A huge number of proteins corresponding to these sequences function through protein-protein binding in cellular systems. Therefore, it is crucial to annotate these entities with predicted protein-protein interactions in both discovery and biological science platforms. Prediction of protein-protein interactions using sequence characteristics with common interface features and molecular mechanics based geometric optimization will be a common practice. We believe that the future provides tremendous opportunity for the application of protein-protein binding knowledge towards the definition of novel protein complexes as candidates toward the prevention and treatment of several life-threatening diseases in the healthcare sector.

Useful Formula

S. No	Parameter	Formula
1	Sensitivity (SE)	$\text{SE} = \frac{TP}{(TP + FN)} * 100.$
2	Specificity (SP)	$\text{SP} = \frac{TN}{(TN + FP)} * 100.$
3	Accuracy (AC)	$\%\text{AC} = \frac{TP + TN}{TP + FP + TN + FN} * 100.$
4	Negative predictive value (NPV)	$\text{NPV} = \frac{TN}{(TN + FN)} * 100.$
5	Positive predictive value (PPV)	$\text{PPV} = \frac{TP}{(TP + FP)} * 100.$
6	TP	True positive
7	TN	True negative
8	FP	False positive
9	FN	False negative

P. Kangueane, C. Nilofer, *Protein-Protein and Domain-Domain Interactions*,
https://doi.org/10.1007/978-981-10-7347-2

The manufacturer's authorised representative in the EU is Springer Nature Customer Service Centre GmbH, Europaplatz 3, 69115 Heidelberg, Germany. If you have any concerns regarding our products, please contact ProductSafety@springernature.com

Printed and bound by CPI Group (UK) Ltd, Croydon, CR0 4YY
01/10/2026
02225866-0001